Power Quick Reference

for the Electrical and Computer PE Exam

John A. Camara, PE

Professional Publications, Inc. • Belmont, California

Benefit by Registering This Book with PPI

- Get book updates and corrections.
- Hear the latest exam news.
- Obtain exclusive exam tips and strategies.
- Receive special discounts.

Register your book at **ppi2pass.com/register**.

Report Errors and View Corrections for This Book

PPI is grateful to every reader who notifies us of a possible error. Your feedback allows us to improve the quality and accuracy of our products. You can report errata and view corrections at **ppi2pass.com/errata**.

POWER QUICK REFERENCE FOR THE ELECTRICAL AND COMPUTER PE EXAM

Current printing of this edition: 2

Printing History

date	edition number	printing number	update
Apr 2016	1	1	New product.
Dec 2016	1	2	Minor corrections. Minor cover updates.

Printed in the United States of America.

PPI
1250 Fifth Avenue
Belmont, CA 94002
(650) 593-9119
ppi2pass.com

ISBN: 978-1-59126-506-1

Library of Congress Control Number: 2015955812

F E D C B A

Table of Contents

How to Use This Book

This book (and others in the *Quick Reference* series) was developed to help you minimize problem-solving time on the PE exam. This *Quick Reference* is a consolidation of the most useful equations, tables, and figures in the *Power Reference Manual*. Using *Quick Reference*, you will not need to wade through pages of descriptive text to find the formulas that remind you of your next solution step.

The idea is this: you study for the exam using (primarily) your *Reference Manual*, and you take the exam using (primarily) your *Quick Reference*.

This book follows the same order, has the same nomenclature, and uses the same chapter, section, equation, figure, and table numbers as the *Power Reference Manual*. Once you become familiar with the sequencing of subjects in the *Reference Manual*, you will be at home with *Quick Reference*. Furthermore, you can always go to the corresponding chapter and section number in the *Reference Manual* for additional information about a subject.

Once you have studied and mastered the theory behind an exam topic, you are ready to tackle the task of solving practice problems. By providing the essential formulas to solve most problems, *Quick Reference* will help you save time solving problems right from the start. In fact, as you progress in your understanding of topics, you will find that you can rely more and more on *Quick Reference* for rapid retrieval of formulas—without needing to refer back to the *Reference Manual*.

When solving problems, you will find that there will be times when you need access to two kinds of information simultaneously: formulas and data, formulas and nomenclature, or formulas and theory. It is likely you have experienced the frustration of having to work problems while substituting a spare pencil, calculator, or shoelace for a page marker in a single book. With *Quick Reference*, you have a convenient way to keep the equations you need in front of you, even as you may be flip-flopping between formulas, theory, and data in your other references.

Quick Reference also provides a convenient place for neatnik engineers to add comments and reminders to equations, without having to mess up a primary reference. I expect you to recycle this book after the exam, so go ahead and write in it.

Once you start incorporating *Quick Reference* into your problem-solving routine, I predict you will not want to return to the one-book approach. *Quick Reference* will save you precious time—and that is how to use this book.

Codes and References

The information that was used to write and update this book was based on the exam specifications at the time of publication. However, as with engineering practice itself, the PE exam is not always based on the most current codes or cutting-edge technology. Similarly, codes, standards, and regulations adopted by state and local agencies often lag issuance by several years. It is likely that the codes that are most current, the codes that you use in practice, and the codes that are the basis of your exam will all be different. However, differences between code editions typically minimally affect the technical accuracy of this book. For more information about the variety of codes related to electrical engineering, refer to the following organizations and their websites.

American National Standards Institute (ansi.org)
Electronic Components Industry Association (ecianow.org)
Federal Communications Commission (fcc.gov)
Institute of Electrical and Electronics Engineers (ieee.org)
International Organization for Standardization (iso.org)
International Society of Automation (isa.org)
National Electrical Manufacturers Association (nema.org)
National Fire Protection Association (nfpa.org)

The PPI website (**ppi2pass.com/eefaq**) provides the dates and editions of the codes, standards, and regulations on which NCEES has announced the PE exams are based. It is your responsibility to find out which codes are relevant to your exam.

CODES AND STANDARDS

47 CFR 73: *Code of Federal Regulations*, "Title 47—Telecommunication, Part 73—Radio Broadcast Rules," 2014. Office of the Federal Register National Archives and Records Administration, Washington, DC. (Communications.)

IEEE/ASTM SI 10: *American National Standard for Metric Practice*, 2010. ASTM International, West Conshohocken, PA. (Metric.)

IEEE Std 141 (IEEE Red Book): *IEEE Recommended Practice for Electric Power Distribution for Industrial Plants*, 1993. The Institute of Electrical and Electronics Engineers, Inc., New York, NY.

IEEE Std 142 (IEEE Green Book): *IEEE Recommended Practice for Grounding of Industrial and Commercial Power Systems*, 2007. The Institute of Electrical and Electronics Engineers, Inc., New York, NY.

IEEE Std 241 (IEEE Gray Book): *IEEE Recommended Practice for Electrical Power Systems in Commercial Buildings*, 1990. The Institute of Electrical and Electronics Engineers, Inc., New York, NY.

IEEE Std 242 (IEEE Buff Book): *IEEE Recommended Practice for Protection and Coordination of Industrial and Commercial Power Systems*, 2001. The Institute of Electrical and Electronics Engineers, Inc., New York, NY.

IEEE Std 399 (IEEE Brown Book): *IEEE Recommended Practice for Industrial and Commercial Power Systems Analysis*, 1997. The Institute of Electrical and Electronics Engineers, Inc., New York, NY.

IEEE Std 446 (IEEE Orange Book): *IEEE Recommended Practice for Emergency and Standby Power Systems for Industrial and Commercial Applications*, 1995. The Institute of Electrical and Electronics Engineers, Inc., New York, NY.

IEEE Std 493 (IEEE Gold Book): *IEEE Recommended Practice for the Design of Reliable Industrial and Commercial Power Systems*, 2007. The Institute of Electrical and Electronics Engineers, Inc., New York, NY.

IEEE Std 551 (IEEE Violet Book): *IEEE Recommended Practice for Calculating Short-Circuit Currents in Industrial and Commercial Power Systems*, 2006. The Institute of Electrical and Electronics Engineers, Inc., New York, NY.

IEEE Std 602 (IEEE White Book): *IEEE Recommended Practice for Electric Systems in Health Care Facilities*, 2007. The Institute of Electrical and Electronics Engineers, Inc., New York, NY.

IEEE Std 739 (IEEE Bronze Book): *IEEE Recommended Practice for Energy Management in Industrial and Commercial Facilities*, 1995. The Institute of Electrical and Electronics Engineers, Inc., New York, NY.

IEEE Std 902 (IEEE Yellow Book): *IEEE Guide for Maintenance, Operation, and Safety of Industrial and Commercial Power Systems*, 1998. The Institute of Electrical and Electronics Engineers, Inc., New York, NY.

IEEE Std 1015 (IEEE Blue Book): *IEEE Recommended Practice for Applying Low-Voltage Circuit Breakers Used in Industrial and Commercial Power Systems*, 2006. The Institute of Electrical and Electronics Engineers, Inc., New York, NY.

IEEE Std 1100 (IEEE Emerald Book): *IEEE Recommended Practice for Powering and Grounding Electronic Equipment*, 2005. The Institute of Electrical and Electronics Engineers, Inc., New York, NY.

NEC (NFPA 70): *National Electrical Code*, 2014. National Fire Protection Association, Quincy, MA. (Power.)

NESC (IEEE C2): *2012 National Electrical Safety Code*, 2012. The Institute of Electrical and Electronics Engineers, Inc., New York, NY. (Power.)

REFERENCES

Anthony, Michael A. *NEC Answers*. New York, NY: McGraw-Hill. (*National Electrical Code* example applications textbook.)

Bronzino, Joseph D. *The Biomedical Engineering Handbook*. Boca Raton, FL: CRC Press. (Electrical and electronics handbook.)

Chemical Rubber Company. *CRC Standard Mathematical Tables and Formulae*. Boca Raton, FL: CRC Press. (General engineering reference.)

Croft, Terrell and Wilford I. Summers. *American Electricians' Handbook*. New York, NY: McGraw-Hill. (Power handbook.)

Earley, Mark W. et al. *National Electrical Code Handbook*, 2014 ed. Quincy, MA: National Fire Protection Association. (Power handbook.)

Fink, Donald G. and H. Wayne Beaty. *Standard Handbook for Electrical Engineers*. New York, NY: McGraw-Hill. (Power and electrical and electronics handbook.)

Grainger, John J. and William D. Stevenson, Jr. *Power System Analysis*. New York, NY: McGraw-Hill. (Power textbook.)

Horowitz, Stanley H. and Arun G. Phadke. *Power System Relaying*. Chichester, West Sussex: John Wiley & Sons, Ltd. (Power protection textbook.)

Huray, Paul G. *Maxwell's Equations*. Hoboken, NJ: John Wiley & Sons, Inc. (Power and electrical and electronics textbook.)

Jaeger, Richard C. and Travis Blalock. *Microelectronic Circuit Design*. New York, NY: McGraw-Hill Education. (Electronic fundamentals textbook.)

Lee, William C.Y. *Wireless and Cellular Telecommunications*. New York, NY: McGraw-Hill. (Electrical and electronics handbook.)

Marne, David J. *National Electrical Safety Code (NESC) 2012 Handbook*. New York, NY: McGraw-Hill Professional. (Power handbook.)

McMillan, Gregory K. and Douglas Considine. *Process/Industrial Instruments and Controls Handbook*. New York, NY: McGraw-Hill Professional. (Power and electrical and electronics handbook.)

Millman, Jacob and Arvin Grabel. *Microelectronics*. New York, NY: McGraw-Hill. (Electronic fundamentals textbook.)

Mitra, Sanjit K. *An Introduction to Digital and Analog Integrated Circuits and Applications*. New York, NY: Harper & Row. (Digital circuit fundamentals textbook.)

Parker, Sybil P., ed. *McGraw-Hill Dictionary of Scientific and Technical Terms*. New York, NY: McGraw-Hill. (General engineering reference.)

Plonus, Martin A. *Applied Electromagnetics*. New York, NY: McGraw-Hill. (Electromagnetic theory textbook.)

Rea, Mark S., ed. *The IESNA Lighting Handbook: Reference & Applications*. New York, NY: Illuminating Engineering Society of North America. (Power handbook.)

Shackelford, James F. and William Alexander, eds. *CRC Materials Science and Engineering Handbook*. Boca Raton, FL: CRC Press, Inc. (General engineering handbook.)

Van Valkenburg, M.E. and B.K. Kinariwala. *Linear Circuits*. Englewood Cliffs, NJ: Prentice-Hall. (AC/DC fundamentals textbook.)

Wildi, Theodore and Perry R. McNeill. *Electrical Power Technology*. New York, NY: John Wiley & Sons. (Power theory and application textbook.)

For Instant Recall

Temperature Conversions

$$°\mathrm{F} = 32° + \tfrac{9}{5}°\mathrm{C}$$
$$°\mathrm{C} = \tfrac{5}{9}(°\mathrm{F} - 32°)$$
$$°\mathrm{R} = °\mathrm{F} + 460°$$
$$\mathrm{K} = °\mathrm{C} + 273°$$
$$\Delta°\mathrm{R} = \tfrac{9}{5}\Delta\mathrm{K}$$
$$\Delta\mathrm{K} = \tfrac{5}{9}\Delta°\mathrm{R}$$

SI Prefixes

symbol	prefix	value
a	atto	10^{-18}
f	femto	10^{-15}
p	pico	10^{-12}
n	nano	10^{-9}
μ	micro	10^{-6}
m	milli	10^{-3}
c	centi	10^{-2}
d	deci	10^{-1}
da	deka	10
h	hecto	10^{2}
k	kilo	10^{3}
M	mega	10^{6}
G	giga	10^{9}
T	tera	10^{12}
P	peta	10^{15}
E	exa	10^{18}

Equivalent Units of Derived and Common SI Units

symbol	equivalent units					
A	C/S	W/V	V/Ω	J/s·V	N/T·m	Wb/H
C	A·s	J/V	N·m/V	V·F	V·s/Ω	W·F/A
F	C/V	C^2/J	s/Ω	A·s/V	$C^2/N\cdot m^3$	$A^2\cdot s^4/kg\cdot m^2$
F/m	C/V·m	$C^2/J\cdot m$	$C^2/N\cdot m^2$	s/Ω·m	A·s/V·m	$A^2\cdot s^4/kg\cdot m^3$
H	Wb/A	V·s/A	Ω·s	$T\cdot m^2/A$	N·s·m/C·A	$kg\cdot m/A^2\cdot s^2$
Hz	1/s	s^{-1}				
J	N·m	V·C	W·s	$kg\cdot m^2/s^2$		
m^2/s^2	J/kg	N·m/kg	V·C/kg	$C\cdot m^2/A\cdot s^3$		
N	J/m	V·C/m	W·C/A·m	$kg\cdot m/s^2$		
N/A^2	$Wb/N\cdot m^2$	$V\cdot s/N\cdot m^2$	T/N	1/A·m		
Pa	N/m^2	J/m^3	$W\cdot s/m^3$	$kg\cdot m/s^2$		
Ω	V/A	W/A^2	V^2/W	$kg\cdot m^2/A^2\cdot s^3$		
S	A/V	1/Ω	A^2/W	$A^2\cdot s^3/A\cdot^2$		
T	Wb/m^2	N/A·m	N·s/C·m	$kg/A\cdot s^2$		
V	J/C	W/A	C/F	$kg\cdot m^2/A\cdot s^3$		
V/m	N/C	W/A·m	J/A·m·s	$kg\cdot m/A\cdot s^3$		
W	J/s	V·A	$V^2/Ω$	$kg\cdot m^2/s^3$		
Wb	V·s	H·A	$T\cdot m^2$	$kg\cdot m^2/A\cdot s^2$		

Fundamental and Physical Constants

quantity	symbol	U.S.	SI
Charge			
electron	e		-1.6022×10^{-19} C
proton	p		$+1.6022 \times 10^{-19}$ C
Density			
air [STP] [32°F (0°C)]		0.0805 lbm/ft^3	1.29 kg/m^3
air [70°F (20°C), 1 atm]		0.0749 lbm/ft^3	1.20 kg/m^3
earth [mean]		345 lbm/ft^3	5520 kg/m^3
mercury		849 lbm/ft^3	1.360×10^4 kg/m^3
seawater		64.0 lbm/ft^3	1025 kg/m^3
water [mean]		62.4 lbm/ft^3	1000 kg/m^3
Distance [mean]			
earth radius		2.09×10^7 ft	6.370×10^6 m
earth-moon separation		1.26×10^9 ft	3.84×10^8 m
earth-sun separation		4.89×10^{11} ft	1.49×10^{11} m
moon radius		5.71×10^6 ft	1.74×10^6 m
sun radius		2.28×10^9 ft	6.96×10^8 m
first Bohr radius	a_0	1.736×10^{-10} ft	5.292×10^{-11} m
Gravitational Acceleration			
earth [mean]	g	32.174 (32.2) ft/sec^2	9.8067 (9.81) m/s^2
moon [mean]		5.47 ft/sec^2	1.67 m/s^2
Mass			
atomic mass unit	u	3.66×10^{-27} lbm	1.6606×10^{-27} kg
earth		1.32×10^{25} lbm	6.00×10^{24} kg
electron [rest]	m_e	2.008×10^{-30} lbm	9.109×10^{-31} kg
moon		1.623×10^{23} lbm	7.36×10^{22} kg
neutron [rest]	m_n	3.693×10^{-27} lbm	1.675×10^{-27} kg
proton [rest]	m_p	3.688×10^{-27} lbm	1.673×10^{-27} kg
sun		4.387×10^{30} lbm	1.99×10^{30} kg
Pressure, atmospheric		14.696 (14.7) lbf/in^2	1.0133×10^5 Pa
Temperature, standard		32°F (492°R)	0°C (273K)
absolute zero		−459.67°F (0°R)	−273.16°C (0K)
triple point, water		32.02°F, 0.0888 psia	0.01109°C, 0.6123 kPa
Velocity			
earth escape		3.67×10^4 ft/sec	1.12×10^4 m/s
light [vacuum]	c	9.84×10^8 ft/sec	$2.9979 (3.00) \times 10^8$ m/s
sound [air, STP]	a	1090 ft/sec	331 m/s
[air, 70°F (20°C)]		1130 ft/sec	344 m/s
Volume, molal ideal gas [STP]		359 ft^3/lbmol	22.41 m^3/kmol
Fundamental Constants			
Avogadro's number	N_A		6.022×10^{23} mol^{-1}
Bohr magneton	μ_B		9.2732×10^{-24} J/T
Boltzmann constant	k	5.65×10^{-24} ft-lbf/°R	1.38065×10^{-23} J/K
Faraday constant	F		96 487 C/mol
gravitational constant	g_c	32.174 lbm-ft/lbf-sec^2	n.a.
gravitational constant	G	3.440×10^{-8} ft^4/lbf-sec^4	6.674×10^{-11} N·m^2/kg^2
nuclear magneton	μ_N		5.050×10^{-27} J/T
permeability of a vacuum	μ_0		1.2566×10^{-6} N/A^2 (H/m)
permittivity of a vacuum	ϵ_0		8.854×10^{-12} C^2/N·m^2 (F/m)
Planck's constant	h		6.6256×10^{-34} J·s
Rydberg constant	R_∞		1.097×10^7 m^{-1}
specific gas constant, air	R	53.35 ft-lbf/lbm-°R	287.03 J/kg·K
Stefan-Boltzmann constant	σ	1.713×10^{-9} Btu/ft^2-hr-°R^4	5.670×10^{-8} W/m^2·K^4
universal gas constant	R^*	1545.35 ft-lbf/lbmol-°R	8314.47 J/kmol·K
	R^*	1.986 Btu/lbmol-°R	0.08206 atm·L/mol·K

Mathematical Constants		
Archimedes' number (pi)	π	3.14159 26536
base of natural logs	e	2.71828 18285
Euler constant	γ	0.57721 56649

Engineering Conversions

(5 significant digits, rounded)

(Atmospheres are standard; Btus are international thermal; calories are gram-calories; gallons are U.S. liquid; horsepower are international; kilocalories are international thermal; miles are statute; pounds are avoirdupois.)

multiply	by	to obtain
acre	43,560	ft^2
Å (angstrom)	1.0×10^{-10}	m
atm	1.01325	bar
atm	76.0	cm Hg (0°C)
atm	33.899	ft H_2O (4°C)
atm	406.783	in H_2O (4°C)
atm	29.921	in Hg (0°C)
atm	101.325	kPa
atm	14.696	lbf/in^2
bar	0.98692	atm
bar	100	kPa
bar	14.504	lbf/in^2
bar	0.1	MPa
Btu	778.17	ft-lbf
Btu	3.9301×10^{-4}	hp-hr
Btu	0.25200	kcal
Btu	1.0551	kJ
Btu	2.9307×10^{-4}	kW-hr
Btu	1.0×10^{-5}	therm
Btu/hr	0.21616	ft-lbf/sec
Btu/hr	3.9301×10^{-4}	hp
Btu/hr	2.9307×10^{-4}	kW
Btu/lbm	2.3260	kJ/kg
Btu/lbm-°R	4.1868	kJ/kg·K
cal (see kcal)	0.001	kcal
cm (see m)	0.01	m
cm (see m)	0.03281	ft
cm (see m)	0.3937	in
cm Hg (0°C)	0.013158	atm
cm^3/g	1.0×10^{-3}	m^3/kg
cP (see P)	0.01	P
eV	1.6022×10^{-22}	kJ
ft	0.30480	m
ft	1.8939×10^{-4}	mi
ft^2	2.2957×10^{-5}	acre
ft H_2O (4°C)	0.029500	atm
ft H_2O (4°C)	0.43353	lbf/in^2
ft^3	7.4805	gal
ft^3/sec	448.83	gal/min
ft-lbf	1.2851×10^{-3}	Btu
ft-lbf	0.0013558	kJ
ft-lbf	3.7661×10^{-7}	kW-hr
ft-lbf	1.3558	N·m
ft-lbf/sec	4.6262	Btu/hr
ft/min	0.00508	m/s
g/cm^3	1000	kg/m^3
g/cm^3	62.428	lbm/ft^3
gal	0.13368	ft^3
gal	3.7854	L
gal	0.0037854	m^3
gal/min	0.0022280	ft^3/sec
gr (grain)	1.4286×10^{-4}	lbm
hp	2544.4	Btu/hr
hp	550	ft-lbf/sec
hp	33,000	ft-lbf/min
hp	0.74570	kW
hp-hr	2544.4	Btu
in	0.02540	m
in H_2O (4°C)	2.4583×10^{-3}	atm
in Hg (0°C)	0.033421	atm
in Hg (0°C)	0.49115	lbf/in^2
J (see kJ)	0.001	kJ
J	9.4782×10^{-4}	Btu
J	6.2415×10^{18}	eV
J	0.73756	ft-lbf
J	1.0	N·m
J/s	1.0	W
kcal	3.9683	Btu
kcal	4.1868	kJ
kg	2.2046	lbm
kg/m^3	0.062428	lbm/ft^3
kip	4.4482	kN
kip	1000	lbf
kJ	0.94782	Btu
kJ	6.2415×10^{21}	eV
kJ	737.56	ft-lbf
kJ	0.23885	kcal
kJ	2.7778×10^{-4}	kW-hr
kJ/kg	0.42992	Btu/lbm
kJ/kg·K	0.23885	Btu/lbm-°R
km	3280.8	ft
km	0.62137	mi
km/hr	0.62137	mi/hr
kN	0.22481	kip
kPa	9.8692×10^{-3}	atm
kPa	0.01	bar
kPa	0.14504	lbf/in^2
kPa	1.4504×10^{-4}	ksi
ksi	6894.8	kPa
kW	3412.1	Btu/hr
kW	0.94782	Btu/sec
kW	737.56	ft-lbf/sec
kW	1.3410	hp
kW-hr	3412.1	Btu
kW-hr	2.6552×10^{6}	ft-lbf
kW-hr	3600.0	kJ
L	0.035135	ft^3
L	61.024	in^3
L	0.26417	gal
L	0.001	m^3
L/s	2.1189	ft^3/min
L/s	15.850	gal/min
lbf	4.4482	N
lbf/in^2	0.068046	atm
lbf/in^2	0.068948	bar
lbf/in^2	2.3067	ft H_2O (4°C)
lbf/in^2	2.0360	in Hg (0°C)
lbf/in^2	6.8948	kPa
lbm	7000	gr (grain)
lbm	0.45359	kg
lbm/ft^3	0.016018	g/cm^3
lbm/ft^3	16.018	kg/m^3
m	1.0×10^{10}	Å (angstrom)
m	3.2808	ft
m	39.370	in
m	1.0×10^{6}	μm (micron)
m^3	264.17	gal
mg/L	8.3454	lbm/MG
m/s	196.85	ft/min
mi	5280	ft
mi	1.6093	km
N	0.22481	lbf
N·m	0.73756	ft-lbf
N·m	1.0	J
P (poise)	1	N·s/m^2
Pa (see kPa)	0.001	kPa
Pa (see kPa)	1.4504×10^{-4}	lbf/in^2
therm	1.0×10^{5}	Btu
ton (cooling)	12,000	Btu/hr
μm (micron)	1.0×10^{-6}	m
W (see kW)	0.001	kW
W (see kW)	3.4121	Btu/hr
W (see kW)	0.73756	ft-lbf/sec
W (see kW)	1.3410×10^{-3}	hp
W (see kW)	1.0	J/s

1 kWh = 3.4095×10^{-12} quad

Formulas of Vector Analysis

	rectangular coordinates	cylindrical coordinates	spherical coordinates
conversion to rectangular coordinates		$x = r\cos\theta$ $y = r\sin\theta$ $z = z$	$x = r\sin\phi\cos\theta$ $y = r\sin\phi\sin\theta$ $z = r\cos\phi$
gradient	$\nabla f = \frac{\partial f}{\partial x}\mathbf{i} + \frac{\partial f}{\partial y}\mathbf{j} + \frac{\partial f}{\partial z}\mathbf{k}$	$\nabla f = \frac{\partial f}{\partial r}\mathbf{r} + \frac{1}{r}\frac{\partial f}{\partial \theta}\boldsymbol{\theta} + \frac{\partial f}{\partial z}\mathbf{k}$	$\nabla f = \frac{\partial f}{\partial r}\mathbf{r} + \frac{1}{r}\frac{\partial f}{\partial \phi}\boldsymbol{\phi} + \frac{1}{r\sin\theta}\frac{\partial f}{\partial \theta}\boldsymbol{\theta}$
divergence	$\nabla \cdot \mathbf{A} = \frac{\partial A_x}{\partial x} + \frac{\partial A_y}{\partial y} + \frac{\partial A_z}{\partial z}$	$\nabla \cdot \mathbf{A} = \frac{1}{r}\frac{\partial (rA_r)}{\partial r} + \frac{1}{r}\frac{\partial A_\theta}{\partial \theta} + \frac{\partial A_z}{\partial z}$	$\nabla \cdot \mathbf{A} = \frac{1}{r^2}\frac{\partial (r^2 A_r)}{\partial r} + \frac{1}{r\sin\phi}\frac{\partial (A_\phi \sin\phi)}{\partial \phi} + \frac{1}{r\sin\phi}\frac{\partial A_\theta}{\partial \theta}$
curl	$\nabla \times \mathbf{A} = \begin{vmatrix} \mathbf{i} & \mathbf{j} & \mathbf{k} \\ \frac{\partial}{\partial x} & \frac{\partial}{\partial y} & \frac{\partial}{\partial z} \\ A_x & A_y & A_z \end{vmatrix}$	$\nabla \times \mathbf{A} = \begin{vmatrix} \frac{1}{r}\mathbf{r} & \boldsymbol{\theta} & \frac{1}{r}\mathbf{k} \\ \frac{\partial}{\partial r} & \frac{\partial}{\partial \theta} & \frac{\partial}{\partial z} \\ A_r & rA_\theta & A_z \end{vmatrix}$	$\nabla \times \mathbf{A} = \begin{vmatrix} \frac{\mathbf{r}}{r^2\sin\theta} & \frac{\boldsymbol{\phi}}{r^2\sin\theta} & \frac{\boldsymbol{\theta}}{r} \\ \frac{\partial}{\partial r} & \frac{\partial}{\partial \phi} & \frac{\partial}{\partial \theta} \\ A_r & rA_\phi & rA_\theta\sin\phi \end{vmatrix}$
Laplacian	$\nabla^2 f = \frac{\partial^2 f}{\partial x^2} + \frac{\partial^2 f}{\partial y^2} + \frac{\partial^2 f}{\partial z^2}$	$\nabla^2 f = \frac{1}{r}\frac{\partial f}{\partial r}\left(r\frac{\partial f}{\partial r}\right) + \frac{1}{r^2}\frac{\partial^2 f}{\partial \theta^2} + \frac{\partial^2 \phi}{\partial z^2}$	$\nabla^2 f = \frac{1}{r^2}\frac{\partial}{\partial r}\left(r^2\frac{\partial f}{\partial r}\right) + \frac{1}{r^2\sin\phi}\frac{\partial}{\partial \phi}\left(\sin\phi\frac{\partial f}{\partial \phi}\right) + \frac{1}{r^2\sin^2\phi}\frac{\partial^2 f}{\partial \theta^2}$

Mensuration of Two-Dimensional Areas

Nomenclature

A	total surface area
b	base
c	chord length
d	distance
h	height
L	length
p	perimeter
r	radius
s	side (edge) length, arc length
θ	vertex angle, in radians
ϕ	central angle, in radians

Circular Sector

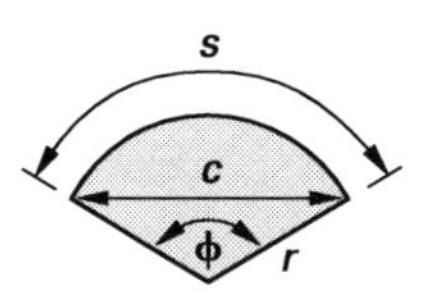

$$A = \tfrac{1}{2}\phi r^2 = \tfrac{1}{2}sr$$

$$\phi = \frac{s}{r}$$

$$s = r\phi$$

$$c = 2r\sin\left(\frac{\phi}{2}\right)$$

Triangle

equilateral

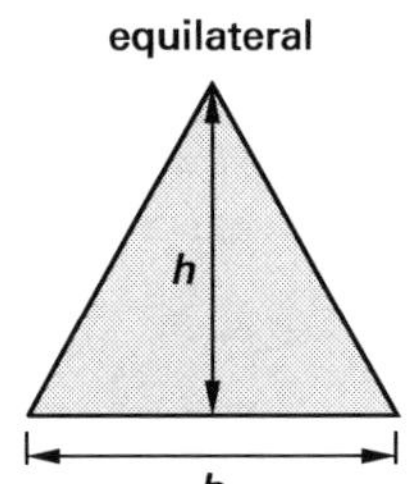

right

H

h

b

oblique

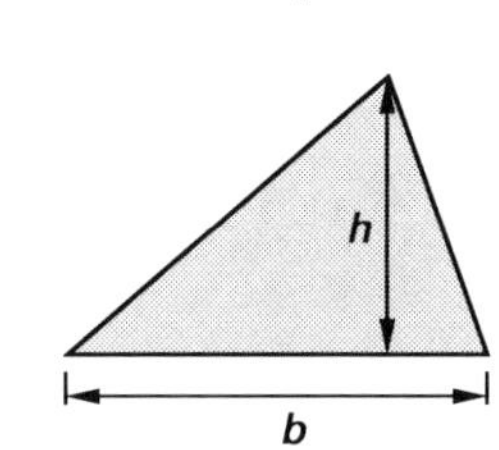

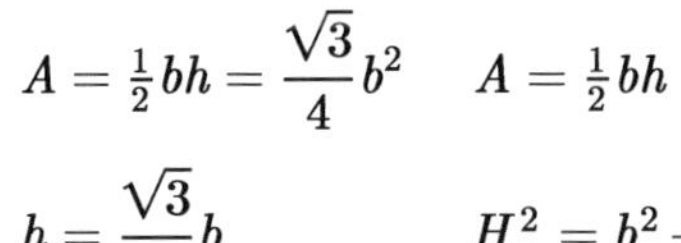

equilateral:

$$A = \tfrac{1}{2}bh = \frac{\sqrt{3}}{4}b^2$$

$$h = \frac{\sqrt{3}}{2}b$$

right:

$$A = \tfrac{1}{2}bh$$

$$H^2 = b^2 + h^2$$

oblique:

$$A = \tfrac{1}{2}bh$$

Parabola

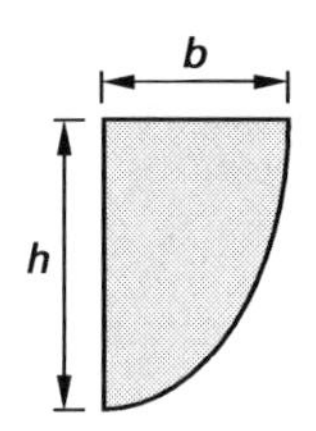

$$A = \tfrac{2}{3}bh$$

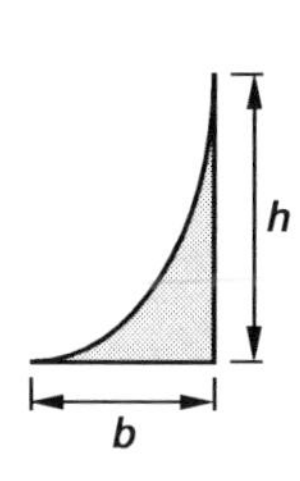

$$A = \tfrac{1}{3}bh$$

Circle

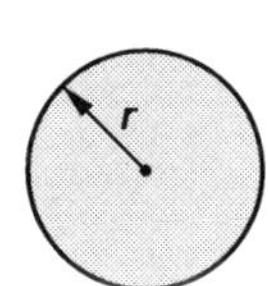

$$p = 2\pi r$$

$$A = \pi r^2 = \frac{p^2}{4\pi}$$

Circular Segment

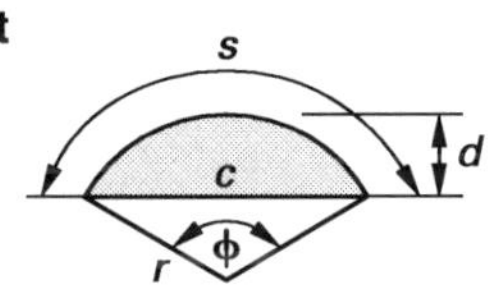

$$A = \tfrac{1}{2}r^2(\phi - \sin\phi)$$

$$\phi = \frac{s}{r} = 2\left(\arccos\frac{r-d}{r}\right)$$

$$c = 2r\sin\left(\frac{\phi}{2}\right)$$

Ellipse

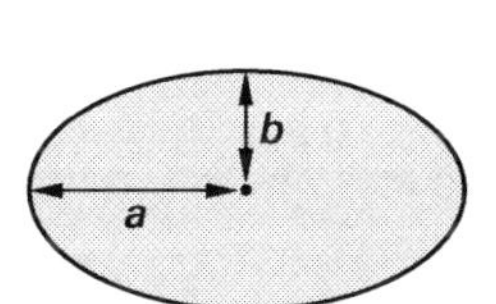

$$A = \pi ab$$

$$p \approx 2\pi\sqrt{\tfrac{1}{2}(a^2 + b^2)} \quad \left[\begin{array}{c}\text{Euler's}\\ \text{upper bound}\end{array}\right]$$

Trapezoid

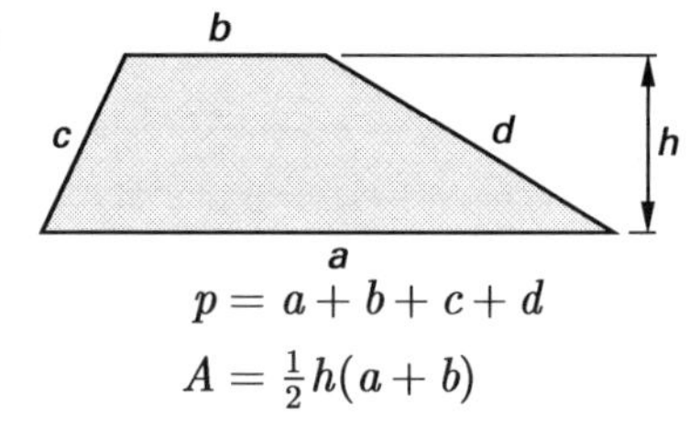

$$p = a + b + c + d$$

$$A = \tfrac{1}{2}h(a+b)$$

If $c = d$, the trapezoid is isosceles.

Regular Polygon (*n* equal sides)

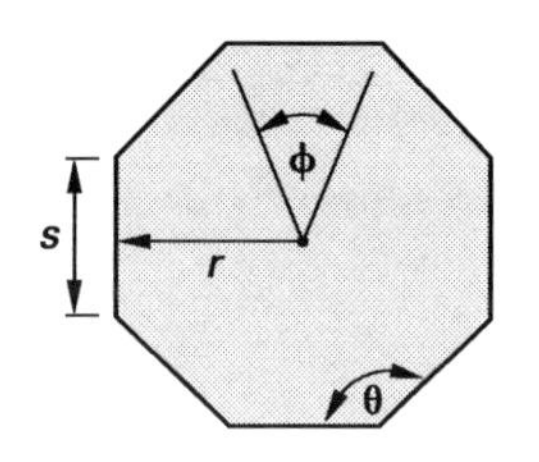

$$\phi = \frac{2\pi}{n}$$

$$\theta = \frac{\pi(n-2)}{n} = \pi - \phi$$

$$p = ns$$

$$s = 2r\tan\frac{\theta}{2}$$

$$A = \tfrac{1}{2}nsr$$

Parallelogram

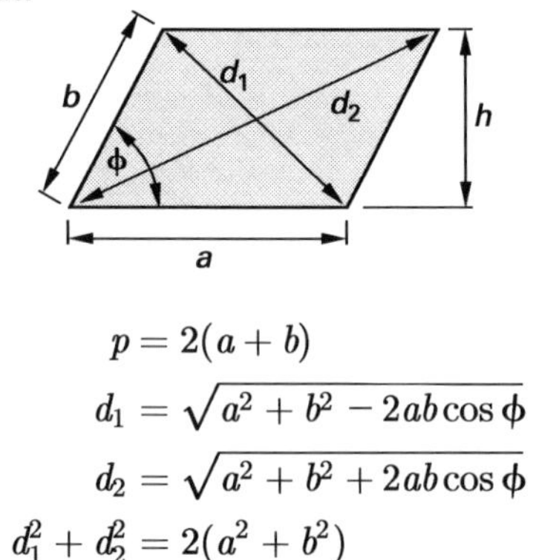

$$p = 2(a+b)$$

$$d_1 = \sqrt{a^2 + b^2 - 2ab\cos\phi}$$

$$d_2 = \sqrt{a^2 + b^2 + 2ab\cos\phi}$$

$$d_1^2 + d_2^2 = 2(a^2 + b^2)$$

$$A = ah = ab\sin\phi$$

If $a = b$, the parallelogram is a rhombus.

regular polygons							
sides	name	area (A) when diameter of inscribed circle = 1	area (A) when side = 1	radius (r) of circumscribed circle when side = 1	length (L) of side when radius (r) of circumscribed circle = 1	length (L) of side when perpendicular to circle = 1	perpendicular (p) to center when side = 1
3	triangle	1.299	0.433	0.577	1.732	3.464	0.289
4	square	1.000	1.000	0.707	1.414	2.000	0.500
5	pentagon	0.908	1.720	0.851	1.176	1.453	0.688
6	hexagon	0.866	2.598	1.000	1.000	1.155	0.866
7	heptagon	0.843	3.634	1.152	0.868	0.963	1.038
8	octagon	0.828	4.828	1.307	0.765	0.828	1.207
9	nonagon	0.819	6.182	1.462	0.684	0.728	1.374
10	decagon	0.812	7.694	1.618	0.618	0.650	1.539
11	undecagon	0.807	9.366	1.775	0.563	0.587	1.703
12	dodecagon	0.804	11.196	1.932	0.518	0.536	1.866

Mensuration of Three-Dimensional Volumes

Nomenclature

A	surface area
b	base
h	height
r	radius
R	radius
s	side (edge) length
V	internal volume

Sphere

$$V = \tfrac{4}{3}\pi r^3 = \tfrac{4}{3}\pi\left(\frac{d}{2}\right)^3 = \tfrac{1}{6}\pi d^3$$

$$A = 4\pi r^2$$

Right Circular Cone (excluding base area)

$$V = \tfrac{1}{3}\pi r^2 h = \tfrac{1}{3}\pi\left(\frac{d}{2}\right)^2 h = \tfrac{1}{12}\pi d^2 h$$

$$A = \pi r\sqrt{r^2 + h^2}$$

Right Circular Cylinder (excluding end areas)

$$V = \pi r^2 h$$

$$A = 2\pi r h$$

Spherical Segment (spherical cap)

Surface area of a spherical segment of radius r cut out by an angle θ_0 rotated from the center about a radius, r, is

$$A = 2\pi r^2(1 - \cos\theta_0)$$

$$\omega = \frac{A}{r^2} = 2\pi(1 - \cos\theta_0)$$

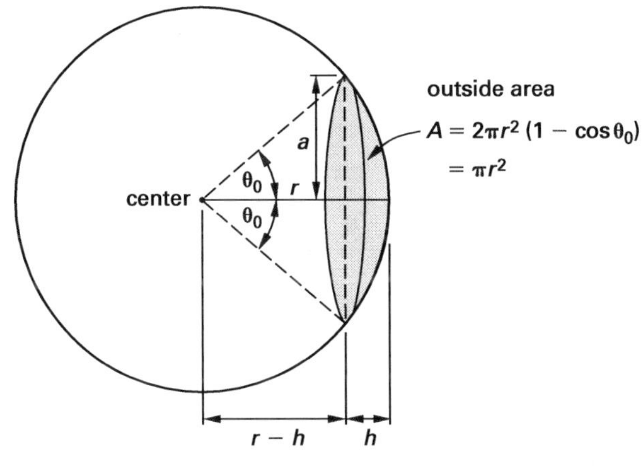

$$V_{cap} = \tfrac{1}{6}\pi h(3a^2 + h^2)$$
$$= \tfrac{1}{3}\pi h^2(3r - h)$$
$$a = \sqrt{h(2r - h)}$$

Paraboloid of Revolution

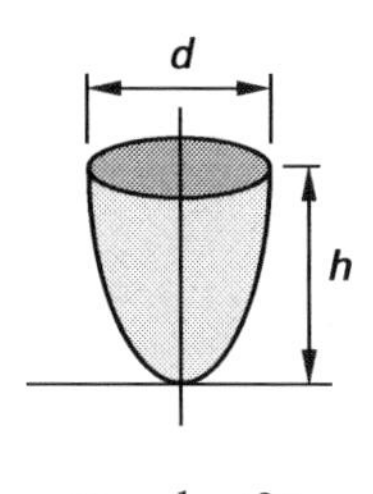

$$V = \tfrac{1}{8}\pi d^2 h$$

Torus

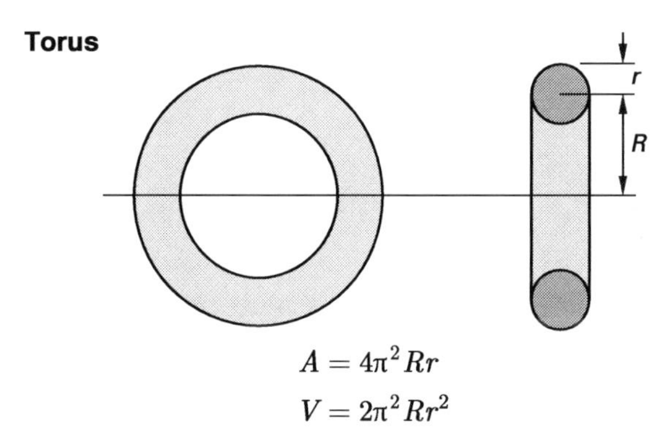

$$A = 4\pi^2 Rr$$

$$V = 2\pi^2 Rr^2$$

Regular Polyhedra (identical faces)

name	number of faces	form of faces	total surface area	volume
tetrahedron	4	equilateral triangle	$1.7321 s^2$	$0.1179 s^3$
cube	6	square	$6.0000 s^2$	$1.0000 s^3$
octahedron	8	equilateral triangle	$3.4641 s^2$	$0.4714 s^3$
dodecahedron	12	regular pentagon	$20.6457 s^2$	$7.6631 s^3$
icosahedron	20	equilateral triangle	$8.6603 s^2$	$2.1817 s^3$

The radius of a sphere inscribed within a regular polyhedron is

$$r = \frac{3 V_{polyhedron}}{A_{polyhedron}}$$

Mathematics

EPRM Chapter 2 Energy, Work, and Power

3. WORK

$$W_{\text{constant force}} = \mathbf{F} \cdot \mathbf{s} = Fs\cos\phi \quad \text{[linear systems]} \qquad 2.6$$

$$W_{\text{constant torque}} = \mathbf{T} \cdot \theta = Fr\theta\cos\phi \quad \text{[rotational systems]} \qquad 2.7$$

Figure 2.1 *Work of a Constant Force*

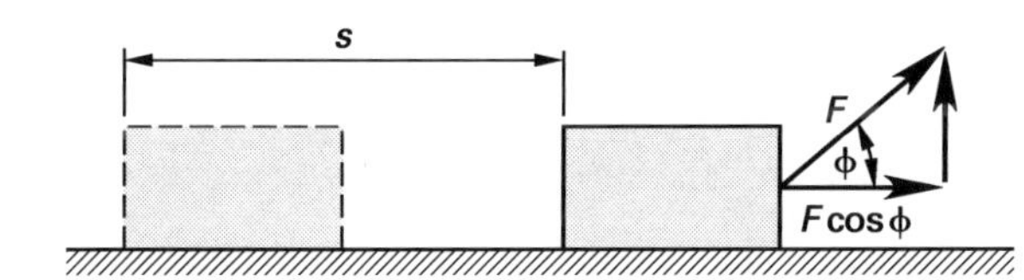

$$W_{\text{friction}} = F_f s \qquad 2.8$$

$$W_{\text{gravity}} = mg(h_2 - h_1) \quad \text{[SI]} \qquad 2.8(a)$$

$$W_{\text{gravity}} = \left(\frac{mg}{g_c}\right)(h_2 - h_1) \quad \text{[U.S.]} \qquad 2.8(b)$$

$$W_{\text{spring}} = \tfrac{1}{2}k\left(\delta_2^2 - \delta_1^2\right) \qquad 2.10$$

4. POTENTIAL ENERGY OF A MASS

$$E_{\text{potential}} = mgh \quad \text{[SI]} \qquad 2.11(a)$$

$$E_{\text{potential}} = \frac{mgh}{g_c} \quad \text{[U.S.]} \qquad 2.11(b)$$

5. KINETIC ENERGY OF A MASS

$$E_{\text{kinetic}} = \tfrac{1}{2}m\mathrm{v}^2 \quad \text{[SI]} \qquad 2.13(a)$$

$$E_{\text{kinetic}} = \frac{m\mathrm{v}^2}{2g_c} \quad \text{[U.S.]} \qquad 2.13(b)$$

$$E_{\text{rotational}} = \tfrac{1}{2}I\omega^2 \quad \text{[SI]} \qquad 2.14(a)$$

$$E_{\text{rotational}} = \frac{I\omega^2}{2g_c} \quad \text{[U.S.]} \qquad 2.14(b)$$

6. SPRING ENERGY

$$E_{\text{spring}} = \tfrac{1}{2}k\delta^2 \qquad 2.16$$

7. PRESSURE ENERGY OF A MASS

$$E_{\text{flow}} = \frac{mp}{\rho} = mpv \quad [v = \text{specific volume}] \qquad 2.17$$

8. INTERNAL ENERGY OF A MASS

$$U_2 - U_1 = Q \qquad 2.18$$

$$Q = mc_v\Delta T \quad \text{[constant-volume process]} \qquad 2.22$$

$$Q = mc_p\Delta T \quad \text{[constant-pressure process]} \qquad 2.23$$

9. WORK-ENERGY PRINCIPLE

$$W = \Delta E = E_2 - E_1 \qquad 2.24$$

11. POWER

$$P = \frac{W}{\Delta t} \qquad 2.25$$

$$P = F\mathrm{v} \quad \text{[linear systems]} \qquad 2.26$$

$$P = T\omega \quad \text{[rotational systems]} \qquad 2.27$$

For flowing fluid,

$$P = \dot{m}\Delta u \qquad 2.28$$

Table 2.2 *Useful Power Conversion Formulas*

1 hp = 550 ft-lbf/sec
= 33,000 ft-lbf/min
= 0.7457 kW
= 0.7068 Btu/sec
1 kW = 737.6 ft-lbf/sec
= 44,250 ft-lbf/min
= 1.341 hp
= 0.9478 Btu/sec
1 Btu/sec = 778.17 ft-lbf/sec
= 46,680 ft-lbf/min
= 1.415 hp

12. EFFICIENCY

$$\eta = \frac{P_{\text{ideal}}}{P_{\text{actual}}} \quad [P_{\text{actual}} \geq P_{\text{ideal}}] \qquad 2.29$$

$$\eta = \frac{P_{\text{actual}}}{P_{\text{ideal}}} \quad [P_{\text{ideal}} \geq P_{\text{actual}}] \qquad 2.30$$

EPRM Chapter 4
Algebra

9. ROOTS OF QUADRATIC EQUATIONS

$$x_1, x_2 = \frac{-b \pm \sqrt{b^2 - 4ac}}{2a} \qquad 4.17$$

13. RULES FOR EXPONENTS AND RADICALS

$$(ab)^n = a^n b^n \qquad 4.24$$

$$b^{m/n} = \sqrt[n]{b^m} = (\sqrt[n]{b})^m \qquad 4.25$$

$$(b^n)^m = b^{nm} \qquad 4.26$$

$$b^m b^n = b^{m+n} \qquad 4.27$$

15. LOGARITHM IDENTITIES

$$\log_b b = 1 \qquad 4.34$$

$$\log_b 1 = 0 \qquad 4.35$$

$$\log_b b^n = n \qquad 4.36$$

$$\log x^a = a \log x \qquad 4.37$$

$$\log xy = \log x + \log y \qquad 4.41$$

$$\log \frac{x}{y} = \log x - \log y \qquad 4.42$$

$$\log_a x = \log_b x \log_a b \qquad 4.43$$

$$\ln x = \ln 10 \log_{10} x \approx 2.3026 \log_{10} x \qquad 4.44$$

$$\log_{10} x = \log_{10} \ln x \, e \approx 0.4343 \ln x \qquad 4.45$$

EPRM Chapter 5
Linear Algebra

5. DETERMINANTS

$$\mathbf{A} = \begin{bmatrix} a & b \\ c & d \end{bmatrix}$$

$$|\mathbf{A}| = \begin{vmatrix} a & b \\ c & d \end{vmatrix} = ad - bc \qquad 5.3$$

$$\mathbf{A} = \begin{bmatrix} a & b & c \\ d & e & f \\ g & h & i \end{bmatrix}$$

$$|\mathbf{A}| = a\begin{vmatrix} e & f \\ h & i \end{vmatrix} - d\begin{vmatrix} b & c \\ h & i \end{vmatrix} + g\begin{vmatrix} b & c \\ e & f \end{vmatrix} \qquad 5.6$$

EPRM Chapter 6
Vectors

2. VECTORS IN *n*-SPACE

$$|\mathbf{V}| = \sqrt{(x_2 - x_1)^2 + (y_2 - y_1)^2} \qquad 6.1$$

$$\phi = \arctan \frac{y_2 - y_1}{x_2 - x_1} \qquad 6.2$$

Figure 6.1 *Vector in Two-Dimensional Space*

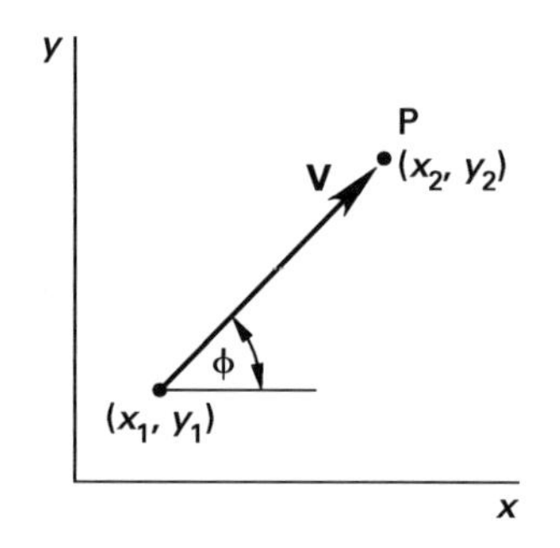

3. UNIT VECTORS

$$\mathbf{V} = |\mathbf{V}|\mathbf{a} = V_x\mathbf{i} + V_y\mathbf{j} + V_z\mathbf{k} \qquad 6.8$$

Figure 6.3 *Cartesian Unit Vectors*

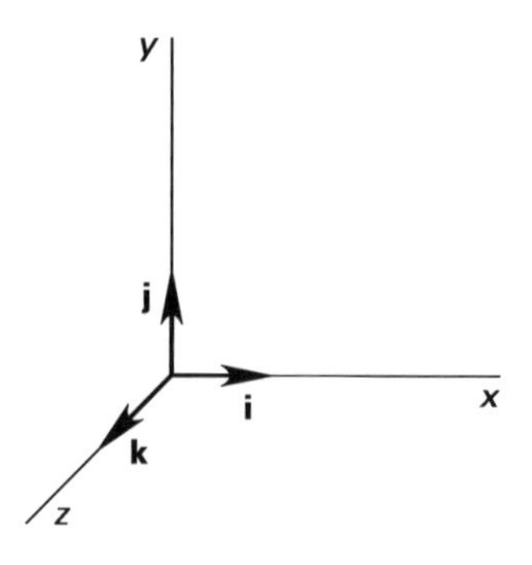

8. VECTOR DOT PRODUCT

$$\mathbf{V}_1 \cdot \mathbf{V}_2 = |\mathbf{V}_1|\,|\mathbf{V}_2|\cos\phi$$

$$= V_{1x}V_{2x} + V_{1y}V_{2y} + V_{1z}V_{2z} \qquad 6.21$$

Figure 6.5 Vector Dot Product

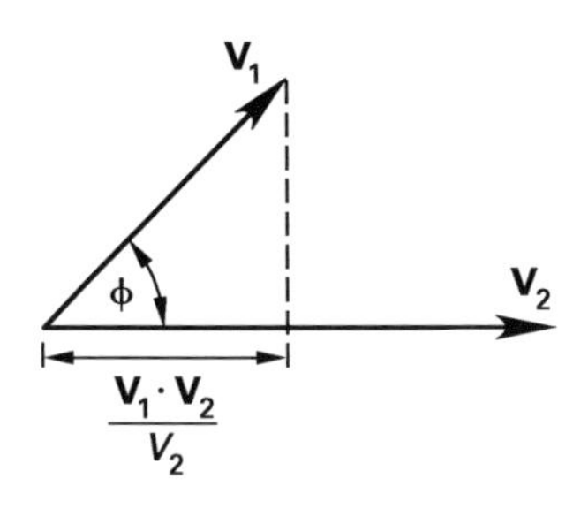

9. VECTOR CROSS PRODUCT

$$|\mathbf{V}_1 \times \mathbf{V}_2| = |\mathbf{V}_1||\mathbf{V}_2|\sin\phi \qquad 6.32$$

Figure 6.6 Vector Cross Product

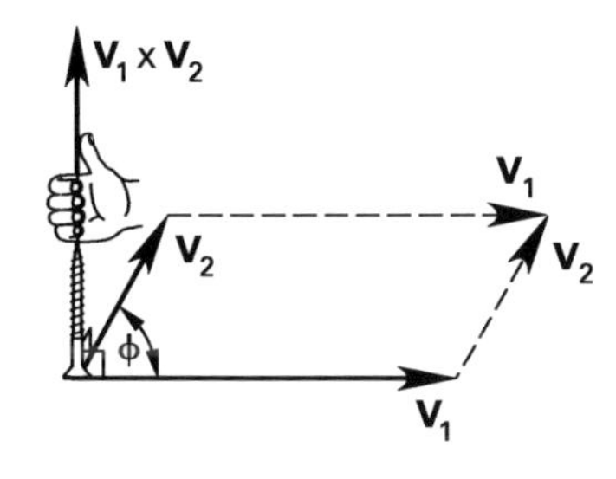

EPRM Chapter 7 Trigonometry

1. DEGREES AND RADIANS

multiply	by	to obtain
radians	$\frac{180}{\pi}$	degrees
degrees	$\frac{\pi}{180}$	radians

4. RIGHT TRIANGLES

$$x^2 + y^2 = r^2 \qquad 7.1$$

5. CIRCULAR TRANSCENDENTAL FUNCTIONS

$$\text{sine: } \sin\theta = \frac{y}{r} = \frac{\text{opposite}}{\text{hypotenuse}} \qquad 7.2$$

$$\text{cosine: } \cos\theta = \frac{x}{r} = \frac{\text{adjacent}}{\text{hypotenuse}} \qquad 7.3$$

$$\text{tangent: } \tan\theta = \frac{y}{x} = \frac{\text{opposite}}{\text{adjacent}} \qquad 7.4$$

$$\text{cotangent: } \cot\theta = \frac{x}{y} = \frac{\text{adjacent}}{\text{opposite}} \qquad 7.5$$

$$\text{secant: } \sec\theta = \frac{r}{x} = \frac{\text{hypotenuse}}{\text{adjacent}} \qquad 7.6$$

$$\text{cosecant: } \csc\theta = \frac{r}{y} = \frac{\text{hypotenuse}}{\text{opposite}} \qquad 7.7$$

$$\cot\theta = \frac{1}{\tan\theta} \qquad 7.8$$

$$\sec\theta = \frac{1}{\cos\theta} \qquad 7.9$$

$$\csc\theta = \frac{1}{\sin\theta} \qquad 7.10$$

Figure 7.6 Trigonometric Functions in a Unit Circle

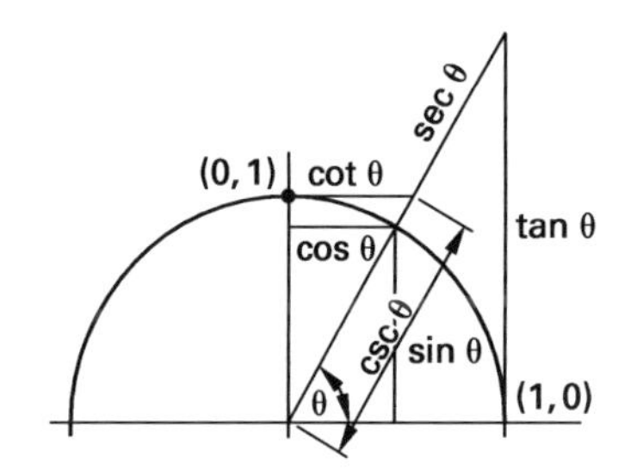

6. SMALL ANGLE APPROXIMATIONS

$$\sin\theta \approx \tan\theta \approx \theta\big|_{\theta < 10^\circ \text{ (0.175 rad)}} \qquad 7.11$$

$$\cos\theta \approx 1\big|_{\theta < 5^\circ \text{ (0.0873 rad)}} \qquad 7.12$$

10. TRIGONOMETRIC IDENTITIES

$$\sin^2\theta + \cos^2\theta = 1 \qquad 7.14$$

$$1 + \tan^2\theta = \sec^2\theta \qquad 7.15$$

$$1 + \cot^2\theta = \csc^2\theta \qquad 7.16$$

- *double-angle formulas*

$$\sin 2\theta = 2\sin\theta\cos\theta = \frac{2\tan\theta}{1+\tan^2\theta} \qquad 7.17$$

$$\cos 2\theta = \cos^2\theta - \sin^2\theta = 1 - 2\sin^2\theta$$

$$= 2\cos^2\theta - 1 = \frac{1-\tan^2\theta}{1+\tan^2\theta} \qquad 7.18$$

$$\tan 2\theta = \frac{2\tan\theta}{1-\tan^2\theta} \qquad 7.19$$

$$\cot 2\theta = \frac{\cot^2\theta - 1}{2\cot\theta} \qquad 7.20$$

- *two-angle formulas*

$$\sin(\theta \pm \phi) = \sin\theta\cos\phi \pm \cos\theta\sin\phi \qquad 7.21$$

$$\cos(\theta \pm \phi) = \cos\theta\cos\phi \mp \sin\theta\sin\phi \qquad 7.22$$

$$\tan(\theta \pm \phi) = \frac{\tan\theta \pm \tan\phi}{1 \mp \tan\theta\tan\phi} \qquad 7.23$$

$$\cot(\theta \pm \phi) = \frac{\cot\phi\cot\theta \mp 1}{\cot\phi \pm \cot\theta} \qquad 7.24$$

- *half-angle formulas* ($\theta < 180°$)

$$\sin\frac{\theta}{2} = \sqrt{\frac{1-\cos\theta}{2}} \qquad 7.25$$

$$\cos\frac{\theta}{2} = \sqrt{\frac{1+\cos\theta}{2}} \qquad 7.26$$

$$\tan\frac{\theta}{2} = \sqrt{\frac{1-\cos\theta}{1+\cos\theta}} = \frac{\sin\theta}{1+\cos\theta} = \frac{1-\cos\theta}{\sin\theta} \qquad 7.27$$

- *miscellaneous formulas* ($\theta < 90°$)

$$\sin\theta = 2\sin\frac{\theta}{2}\cos\frac{\theta}{2} \qquad 7.28$$

$$\sin\theta = \sqrt{\frac{1-\cos 2\theta}{2}} \qquad 7.29$$

$$\cos\theta = \cos^2\frac{\theta}{2} - \sin^2\frac{\theta}{2} \qquad 7.30$$

$$\cos\theta = \sqrt{\frac{1+\cos 2\theta}{2}} \qquad 7.31$$

$$\tan\theta = \frac{2\tan\frac{\theta}{2}}{1-\tan^2\frac{\theta}{2}} = \frac{2\sin\frac{\theta}{2}\cos\frac{\theta}{2}}{\cos^2\frac{\theta}{2} - \sin^2\frac{\theta}{2}} \qquad 7.32$$

$$\tan\theta = \sqrt{\frac{1-\cos 2\theta}{1+\cos 2\theta}} = \frac{\sin 2\theta}{1+\cos 2\theta} = \frac{1-\cos 2\theta}{\sin 2\theta} \qquad 7.33$$

$$\cot\theta = \frac{\cot^2\frac{\theta}{2} - 1}{2\cot\frac{\theta}{2}} = \frac{\cos^2\frac{\theta}{2} - \sin^2\frac{\theta}{2}}{2\sin\frac{\theta}{2}\cos\frac{\theta}{2}} \qquad 7.34$$

$$\cot\theta = \sqrt{\frac{1+\cos 2\theta}{1-\cos 2\theta}} = \frac{1+\cos 2\theta}{\sin 2\theta} = \frac{\sin 2\theta}{1-\cos 2\theta} \qquad 7.35$$

14. GENERAL TRIANGLES

$$\frac{\sin A}{a} = \frac{\sin B}{b} = \frac{\sin C}{c} \qquad 7.57$$

$$a^2 = b^2 + c^2 - 2bc\cos A \qquad 7.58$$

Figure 7.9 General Triangle

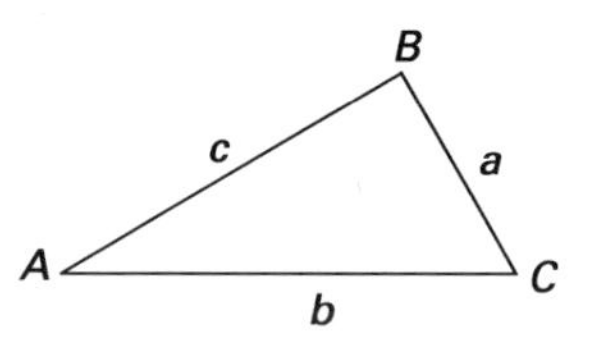

EPRM Chapter 8 Analytic Geometry

2. AREAS WITH IRREGULAR BOUNDARIES

- *trapezoidal rule*

$$A = \frac{d}{2}\left(h_0 + h_n + 2\sum_{i=1}^{n-1} h_i\right) \qquad 8.1$$

- *Simpson's rule*

$$A = \frac{d}{3}\left(h_0 + h_n + 4\sum_{\substack{i\text{ odd}\\ i=1}}^{n-1} h_i + 2\sum_{\substack{i\text{ even}\\ i=2}}^{n-2} h_i\right) \qquad 8.2$$

Figure 8.1 Irregular Areas

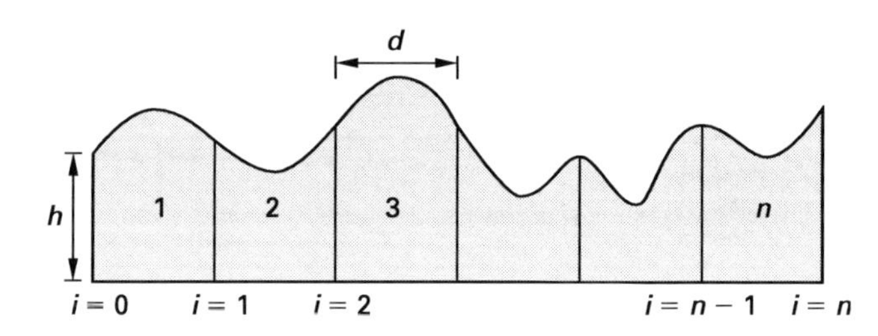

10. STRAIGHT LINES

- *general form*

$$Ax + By + C = 0 \qquad 8.8$$

$$A = -mB \qquad 8.9$$

$$B = \frac{-C}{b} \qquad 8.10$$

$$C = -aA = -bB \qquad 8.11$$

- *slope-intercept form*

$$y = mx + b \qquad 8.12$$

$$m = \frac{-A}{B} = \tan\theta = \frac{y_2 - y_1}{x_2 - x_1} \qquad 8.13$$

$$b = \frac{-C}{B} \qquad 8.14$$

$$a = \frac{-C}{A} \qquad 8.15$$

- *point-slope form*

$$y - y_1 = m(x - x_1) \qquad 8.16$$

- *intercept form*

$$\frac{x}{a} + \frac{y}{b} = 1 \qquad 8.17$$

- *two-point form*

$$\frac{y - y_1}{x - x_1} = \frac{y_2 - y_1}{x_2 - x_1} \qquad 8.18$$

- *normal form*

$$x \cos \beta + y \sin \beta - d = 0 \qquad 8.19$$

(d and β are constants; x and y are variables.)

- *polar form*

$$r = \frac{d}{\cos(\beta - \alpha)} \qquad 8.20$$

(d and β are constants; r and α are variables.)

Figure 8.8 *Straight Line*

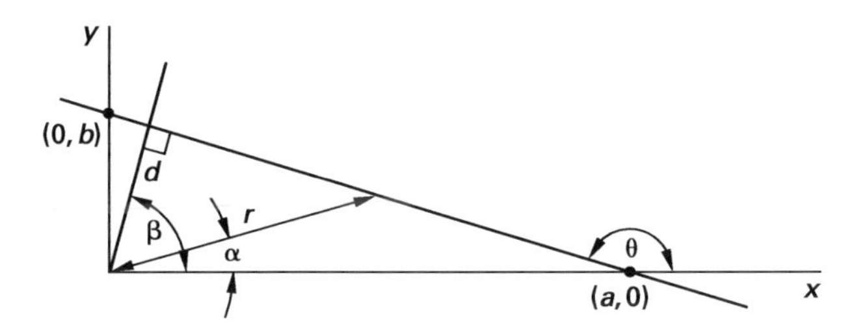

14. DISTANCES BETWEEN GEOMETRIC FIGURES

- between two points in (x, y, z) format:

$$d = \sqrt{(x_2 - x_1)^2 + (y_2 - y_1)^2 + (z_2 - z_1)^2} \qquad 8.48$$

- between a point (x_0, y_0) and a line $Ax + By + C = 0$:

$$d = \frac{|Ax_0 + By_0 + C|}{\sqrt{A^2 + B^2}} \qquad 8.49$$

- between a point (x_0, y_0, z_0) and a plane $Ax + By + Cz + D = 0$:

$$d = \frac{|Ax_0 + By_0 + Cz_0 + D|}{\sqrt{A^2 + B^2 + C^2}} \qquad 8.50$$

- between two parallel lines $Ax + By + C = 0$:

$$d = \left| \frac{|C_2|}{\sqrt{A_2^2 + B_2^2}} - \frac{|C_1|}{\sqrt{A_1^2 + B_1^2}} \right| \qquad 8.51$$

17. CIRCLE

$$Ax^2 + Ay^2 + Dx + Ey + F = 0 \qquad 8.66$$

$$(x - h)^2 + (y - k)^2 = r^2 \qquad 8.67$$

18. PARABOLA

$$(y - k)^2 = 4p(x - h)\Big|_{\text{opens horizontally}} \qquad 8.73$$

$$y^2 = 4px\Big|_{\substack{\text{vertex at origin} \\ h = k = 0}} \qquad 8.74$$

$$(x - h)^2 = 4p(y - k)\big|_{\text{opens vertically}} \qquad 8.75$$

$$x^2 = 4py\big|_{\text{vertex at origin}} \qquad 8.76$$

Figure 8.13 *Parabola*

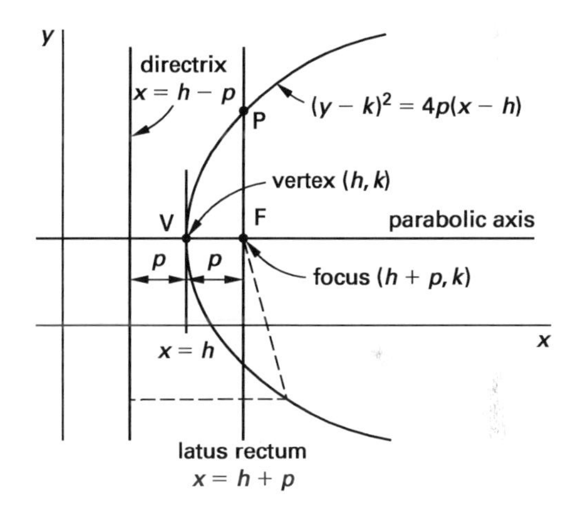

20. HYPERBOLA

Figure 8.15 *Hyperbola*

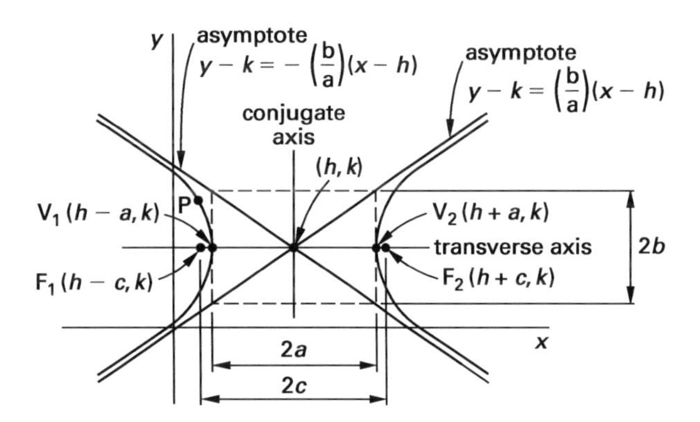

EPRM Chapter 11
Differential Equations

2. HOMOGENEOUS, FIRST-ORDER LINEAR DIFFERENTIAL EQUATIONS WITH CONSTANT COEFFICIENTS

- *general form*

$$y' + ky = 0 \quad 11.1$$

- *solution*

$$y = Ae^{rx} = Ae^{-kx} \quad 11.2$$

3. FIRST-ORDER LINEAR DIFFERENTIAL EQUATIONS

$$y' + p(x)y = g(x) \quad 11.4$$

- *integrating factor*

$$u(x) = \exp\left(\int p(x)dx\right) \quad 11.5$$

- *closed-form solution*

$$y = \frac{1}{u(x)}\left(\int u(x)g(x)dx + C\right) \quad 11.6$$

6. HOMOGENEOUS, SECOND-ORDER LINEAR DIFFERENTIAL EQUATIONS WITH CONSTANT COEFFICIENTS

$$y'' + k_1y' + k_2y = 0 \quad 11.13$$

$$r^2 + k_1r + k_2 = 0 \quad 11.14$$

- *solution if the two roots of Eq. 11.14 are real and different*

$$y = A_1e^{r_1x} + A_2e^{r_2x} \quad 11.15$$

- *solution if the two roots are real and the same*

$$y = A_1e^{rx} + A_2xe^{rx} \quad 11.16$$

$$r = \frac{-k_1}{2} \quad 11.17$$

- *solution if the two roots are imaginary* (they will be of the form $(\alpha + i\omega)$ and $(\alpha - i\omega)$)

$$y = A_1e^{\alpha x}\cos\omega x + A_2e^{\alpha x}\sin\omega x \quad 11.18$$

In all three cases, A_1 and A_2 must be found from the two initial conditions.

7. NONHOMOGENEOUS DIFFERENTIAL EQUATIONS

- *forcing function, $f(x)$*

$$y'' + p(x)y' + q(x)y = f(x) \quad 11.19$$

The *complementary solution*, y_c, solves the complementary (i.e., homogeneous) problem. The *particular solution*, y_p, is any specific solution to the nonhomogeneous Eq. 11.19 that is known or can be found.

$$y = y_c + y_p \quad 11.20$$

Table 11.1 *Particular Solutions**

form of $f(x)$	form of y_p
$P_n(x) = a_0x^n + a_1x^{n-1} + \cdots + a_n$	$x^s\left(A_0x^n + A_1x^{n-1} + \cdots + A_n\right)$
$P_n(x)e^{\alpha x}$	$x^s\left(A_0x^n + A_1x^{n-1} + \cdots + A_n\right)e^{\alpha x}$
$P_n(x)e^{\alpha x}\begin{Bmatrix}\sin\omega x\\ \cos\omega x\end{Bmatrix}$	$x^s\left(\left(A_0x^n + A_1x^{n-1} + \cdots + A_n\right)\times(e^{\alpha x}\cos\omega x) + \left(B_0x^n + B_1x^{n-1} + \cdots + B_n\right)\times(e^{\alpha x}\sin\omega x)\right)$

*$P_n(x)$ is a polynomial of degree n.

9. LAPLACE TRANSFORMS

$$\mathcal{L}(f(t)) = F(s) = \int_0^\infty e^{-st}f(t)dt \quad 11.26$$

11. ALGEBRA OF LAPLACE TRANSFORMS

- *linearity theorem* (c is a constant)

$$\mathcal{L}(cf(t)) = c\mathcal{L}(f(t)) = cF(s) \quad 11.28$$

- *superposition theorem* ($f(t)$ and $g(t)$ are different functions)

$$\mathcal{L}(f(t) \pm g(t)) = \mathcal{L}(f(t)) \pm \mathcal{L}(g(t)) = F(s) \pm G(s) \quad 11.29$$

- *time-shifting theorem* (*delay theorem*)

$$\mathcal{L}(f(t-b)u_b) = e^{-bs}F(s) \quad 11.30$$

- *Laplace transform of a derivative*

$$\mathcal{L}(f^n(t)) = -f^{n-1}(0) - sf^{n-2}(0) - \cdots - s^{n-1}f(0) + s^nF(s) \quad 11.31$$

- *other properties*

$$\mathcal{L}\left(\int_0^t f(u)du\right) = \frac{1}{s}F(s) \quad 11.32$$

$$\mathcal{L}(tf(t)) = -\frac{dF}{ds} \quad 11.33$$

$$\mathcal{L}\left(\frac{1}{t}f(t)\right) = \int_0^\infty F(u)du \quad 11.34$$

12. CONVOLUTION INTEGRAL

$$f(t) = \mathcal{L}^{-1}\big(F_1(s)F_2(s)\big) = \int_0^t f_1(t-\chi)f_2(\chi)d\chi = \int_0^t f_1(\chi)f_2(t-\chi)d\chi \qquad 11.35$$

EPRM Chapter 12
Probability and Statistical Analysis of Data

9. BINOMIAL DISTRIBUTION

$$p\{x\} = f(x) = \binom{n}{x}\hat{p}^x\hat{q}^{n-x} \qquad 12.28$$

$$\binom{n}{x} = \frac{n!}{(n-x)!x!} \qquad 12.29$$

12. POISSON DISTRIBUTION

$$p\{x\} = f(x) = \frac{e^{-\lambda}\lambda^x}{x!} \qquad [\lambda > 0] \qquad 12.34$$

λ is both the mean and the variance of the Poisson distribution.

15. NORMAL DISTRIBUTION

(See the "Areas Under the Standard Normal Curve" table.)

$$z = \frac{x_0 - \mu}{\sigma} \qquad 12.43$$

16. APPLICATION: RELIABILITY

$$R\{t\} = e^{-\lambda t} = e^{-t/\text{MTTF}} \qquad 12.47$$

$$\lambda = \frac{1}{\text{MTTF}} \qquad 12.48$$

$$R\{t\} = 1 - F(t) = 1 - \left(1 - e^{-\lambda t}\right) = e^{-\lambda t} \qquad 12.49$$

$$z\{t\} = \lambda \qquad 12.50$$

18. MEASURES OF CENTRAL TENDENCY

$$\overline{x} = \left(\frac{1}{n}\right)(x_1 + x_2 + \cdots + x_n) = \frac{\sum x_i}{n} \qquad 12.56$$

$$\text{geometric mean} = \sqrt[n]{x_1x_2x_3\cdots x_n} \qquad [x_i > 0] \qquad 12.57$$

$$\text{harmonic mean} = \frac{n}{\dfrac{1}{x_1} + \dfrac{1}{x_2} + \cdots + \dfrac{1}{x_n}} \qquad 12.58$$

$$x_{\text{rms}} = \sqrt{\frac{\sum x_i^2}{n}} \qquad 12.59$$

19. MEASURES OF DISPERSION

$$\sigma = \sqrt{\frac{\sum(x_i - \mu)^2}{N}} = \sqrt{\frac{\sum x_i^2}{N} - \mu^2} \qquad 12.60$$

$$s = \sqrt{\frac{\sum(x_i - \overline{x})^2}{n-1}} = \sqrt{\frac{\sum x_i^2 - \dfrac{\left(\sum x_i\right)^2}{n}}{n-1}} \qquad 12.61$$

$$\sigma_{\text{sample}} = s\sqrt{\frac{n-1}{n}} \qquad 12.62$$

$$\text{coefficient of variation} = \frac{s}{\overline{x}} \qquad 12.63$$

22. APPLICATION: CONFIDENCE LIMITS

$$\text{LCL}: \mu - z_c\sigma$$

$$\text{UCL}: \mu + z_c\sigma$$

$$p\{\overline{x} > L\} = p\left\{z > \left|\frac{L - \mu}{\frac{\sigma}{\sqrt{n}}}\right|\right\} \qquad 12.67$$

$$p\{\text{LCL} < \overline{x} < \text{UCL}\} = p\left\{\frac{\text{LCL} - \mu}{\frac{\sigma}{\sqrt{n}}} < z < \frac{\text{UCL} - \mu}{\frac{\sigma}{\sqrt{n}}}\right\} \qquad 12.68$$

Table 12.1 *Values of z for Various Confidence Levels*

confidence level, C	one-tail limit, z	two-tail limit, z
90%	1.28	1.645
95%	1.645	1.96
97.5%	1.96	2.17
99%	2.33	2.575
99.5%	2.575	2.81
99.75%	2.81	3.00

Areas Under the Standard Normal Curve

(0 to z)

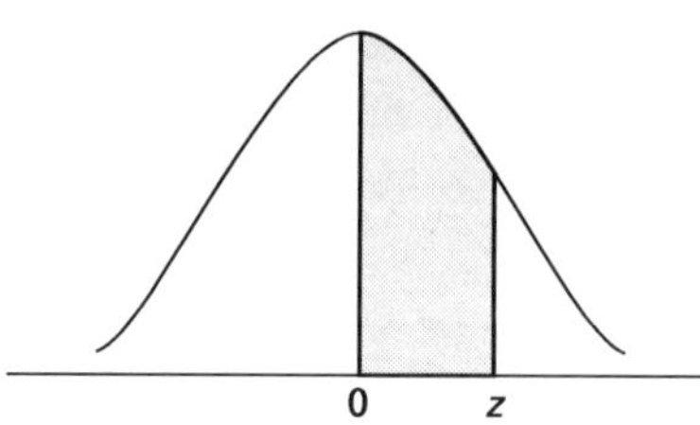

z	0	1	2	3	4	5	6	7	8	9
0.0	0.0000	0.0040	0.0080	0.0120	0.0160	0.0199	0.0239	0.0279	0.0319	0.0359
0.1	0.0398	0.0438	0.0478	0.0517	0.0557	0.0596	0.0636	0.0675	0.0714	0.0754
0.2	0.0793	0.0832	0.0871	0.0910	0.0948	0.0987	0.1026	0.1064	0.1103	0.1141
0.3	0.1179	0.1217	0.1255	0.1293	0.1331	0.1368	0.1406	0.1443	0.1480	0.1517
0.4	0.1554	0.1591	0.1628	0.1664	0.1700	0.1736	0.1772	0.1808	0.1844	0.1879
0.5	0.1915	0.1950	0.1985	0.2019	0.2054	0.2088	0.2123	0.2157	0.2190	0.2224
0.6	0.2258	0.2291	0.2324	0.2357	0.2389	0.2422	0.2454	0.2486	0.2518	0.2549
0.7	0.2580	0.2612	0.2642	0.2673	0.2704	0.2734	0.2764	0.2794	0.2823	0.2852
0.8	0.2881	0.2910	0.2939	0.2967	0.2996	0.3023	0.3051	0.3078	0.3106	0.3133
0.9	0.3159	0.3186	0.3212	0.3238	0.3264	0.3289	0.3315	0.3340	0.3365	0.3389
1.0	0.3413	0.3438	0.3461	0.3485	0.3508	0.3531	0.3554	0.3577	0.3599	0.3621
1.1	0.3643	0.3665	0.3686	0.3708	0.3729	0.3749	0.3770	0.3790	0.3810	0.3830
1.2	0.3849	0.3869	0.3888	0.3907	0.3925	0.3944	0.3962	0.3980	0.3997	0.4015
1.3	0.4032	0.4049	0.4066	0.4082	0.4099	0.4115	0.4131	0.4147	0.4162	0.4177
1.4	0.4192	0.4207	0.4222	0.4236	0.4251	0.4265	0.4279	0.4292	0.4306	0.4319
1.5	0.4332	0.4345	0.4357	0.4370	0.4382	0.4394	0.4406	0.4418	0.4429	0.4441
1.6	0.4452	0.4463	0.4474	0.4484	0.4495	0.4505	0.4515	0.4525	0.4535	0.4545
1.7	0.4554	0.4564	0.4573	0.4582	0.4591	0.4599	0.4608	0.4616	0.4625	0.4633
1.8	0.4641	0.4649	0.4656	0.4664	0.4671	0.4678	0.4686	0.4693	0.4699	0.4706
1.9	0.4713	0.4719	0.4726	0.4732	0.4738	0.4744	0.4750	0.4756	0.4761	0.4767
2.0	0.4772	0.4778	0.4783	0.4788	0.4793	0.4798	0.4803	0.4808	0.4812	0.4817
2.1	0.4821	0.4826	0.4830	0.4834	0.4838	0.4842	0.4846	0.4850	0.4854	0.4857
2.2	0.4861	0.4864	0.4868	0.4871	0.4875	0.4878	0.4881	0.4884	0.4887	0.4890
2.3	0.4893	0.4896	0.4898	0.4901	0.4904	0.4906	0.4909	0.4911	0.4913	0.4916
2.4	0.4918	0.4920	0.4922	0.4925	0.4927	0.4929	0.4931	0.4932	0.4934	0.4936
2.5	0.4938	0.4940	0.4941	0.4943	0.4945	0.4946	0.4948	0.4949	0.4951	0.4952
2.6	0.4953	0.4955	0.4956	0.4957	0.4959	0.4960	0.4961	0.4962	0.4963	0.4964
2.7	0.4965	0.4966	0.4967	0.4968	0.4969	0.4970	0.4971	0.4972	0.4973	0.4974
2.8	0.4974	0.4975	0.4976	0.4977	0.4977	0.4978	0.4979	0.4979	0.4980	0.4981
2.9	0.4981	0.4982	0.4982	0.4983	0.4984	0.4984	0.4985	0.4985	0.4986	0.4986
3.0	0.4987	0.4987	0.4987	0.4988	0.4988	0.4989	0.4989	0.4989	0.4990	0.4990
3.1	0.4990	0.4991	0.4991	0.4991	0.4992	0.4992	0.4992	0.4992	0.4993	0.4993
3.2	0.4993	0.4993	0.4994	0.4994	0.4994	0.4994	0.4994	0.4995	0.4995	0.4995
3.3	0.4995	0.4995	0.4996	0.4996	0.4996	0.4996	0.4996	0.4996	0.4996	0.4997
3.4	0.4997	0.4997	0.4997	0.4997	0.4997	0.4997	0.4997	0.4997	0.4997	0.4998
3.5	0.4998	0.4998	0.4998	0.4998	0.4998	0.4998	0.4998	0.4998	0.4998	0.4998
3.6	0.4998	0.4998	0.4999	0.4999	0.4999	0.4999	0.4999	0.4999	0.4999	0.4999
3.7	0.4999	0.4999	0.4999	0.4999	0.4999	0.4999	0.4999	0.4999	0.4999	0.4999
3.8	0.4999	0.4999	0.4999	0.4999	0.4999	0.4999	0.4999	0.4999	0.4999	0.4999
3.9	0.5000	0.5000	0.5000	0.5000	0.5000	0.5000	0.5000	0.5000	0.5000	0.5000

Basic Theory

EPRM Chapter 16
Electromagnetic Theory

7. COULOMB'S LAW

$$\mathbf{F}_{1\text{-}2} = \frac{Q_1 Q_2}{4\pi\epsilon r_{1\text{-}2}^2}\mathbf{a}_{r_{1-2}} \qquad 16.3$$

$$\mathbf{F}_{1\text{-}2} = \mathbf{F}_2 = Q_2\mathbf{E}_1 \qquad 16.4$$

- *Gaussian SI form*

$$\mathbf{F}_{1\text{-}2} = \frac{kQ_1 Q_2}{\epsilon_r r^2}\mathbf{a}_{1\text{-}2} \qquad 16.5$$

ϵ_r represents the *relative permittivity*. k has an approximate value of 8.987×10^9 N·m^2/C^2.

8. ELECTRIC FIELDS

$$\mathbf{E} = \frac{Q}{4\pi\epsilon r^2}\mathbf{a} \qquad 16.6$$

$$\mathbf{E}_1 = \sum_{i=1}^{i}\frac{Q_i}{4\pi\epsilon r_{i1}^2}\mathbf{a}_{r_{i1}} \qquad 16.7$$

$$\mathbf{E}_1 = \int_{V_{\text{volume}}}\frac{\rho dV_{\text{volume}}}{4\pi\epsilon r_{i1}^2}\mathbf{a}_{r_{i1}} \qquad 16.8$$

The total charge, Q, in the volume V_{volume} that has the charge density ρ is

$$Q = \int_{V_{\text{volume}}}\rho dV_{\text{volume}} \qquad 16.9$$

$$E_{\text{uniform}} = \frac{V_{\text{plates}}}{r} \qquad 16.10$$

9. PERMITTIVITY AND SUSCEPTIBILITY

$$\epsilon = \epsilon_0\epsilon_r \qquad 16.11$$

$$\epsilon_r = \frac{C_{\text{with dielectric}}}{C_{\text{vacuum}}} \qquad 16.12$$

$$\epsilon = \epsilon_0(1 + \chi_e) \qquad 16.13$$

χ_e represents the electric susceptibility.

11. ELECTRIC FLUX DENSITY

$$\mathbf{D} = \epsilon\mathbf{E} = \epsilon_0\epsilon_r\mathbf{E} \qquad 16.15$$

$$D = |\mathbf{D}| = \frac{\psi}{A} = \frac{Q}{A} = \sigma \qquad 16.16$$

12. GAUSS' LAW FOR ELECTROSTATICS

$$\Psi = \iint DdA = \iint \sigma dA = Q \quad [d\mathbf{A} \text{ parallel to } \mathbf{D}] \qquad 16.17$$

$$\Psi = \iint \mathbf{D}\cdot ds = \iint D\cos\theta ds = Q \quad [\text{arbitrary surface}] \qquad 16.18$$

$$\Psi = \iiint \rho dv = Q \quad [\text{arbitrary volume}] \qquad 16.19$$

$$\Psi = \iint \epsilon EdA = Q \quad [d\mathbf{A} \text{ parallel to } \mathbf{E}] \qquad 16.20$$

13. CAPACITANCE AND ELASTANCE

$$Q = CV \qquad 16.22(a)$$

$$C = \frac{Q}{V} \qquad 16.22(b)$$

$$S = \frac{1}{C} = \frac{V}{Q} = \frac{\frac{U}{Q}}{Q} \qquad 16.23$$

14. CAPACITORS

- *parallel plate capacitor*

$$C = \frac{\epsilon A}{r} \qquad 16.24$$

- *total energy, U (in joules)*

$$U = \tfrac{1}{2}CV^2 = \tfrac{1}{2}VQ = \tfrac{1}{2}\frac{Q^2}{C} \qquad 16.25$$

Table 16.6 *Electric Fields and Capacitance for Various Configurations*

isolated point charge 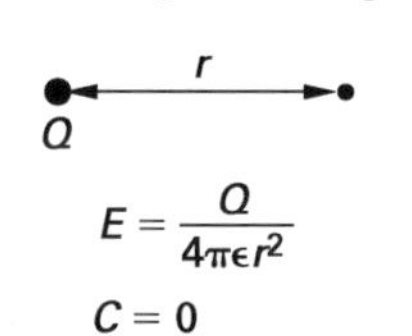 $E=\frac{Q}{4\pi\epsilon r^2}$ $C=0$	infinite coaxial cylinder 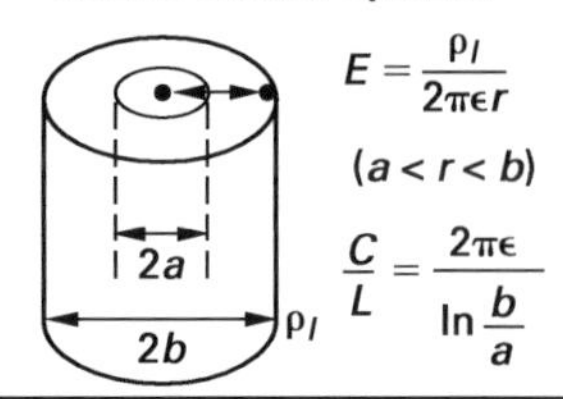 $E=\frac{\rho_l}{2\pi\epsilon r}$ $(a<r<b)$ $\frac{C}{L}=\frac{2\pi\epsilon}{\ln\frac{b}{a}}$
isolated sphere 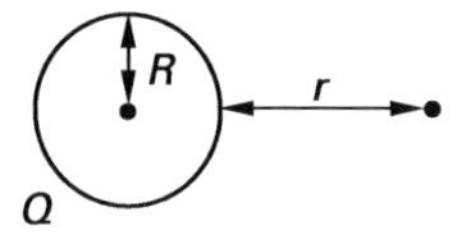 $E=\frac{Q}{4\pi\epsilon(r+R)^2}\quad(r>0)$ $C=4\pi\epsilon R$	infinite line distribution inside an infinite cylinder 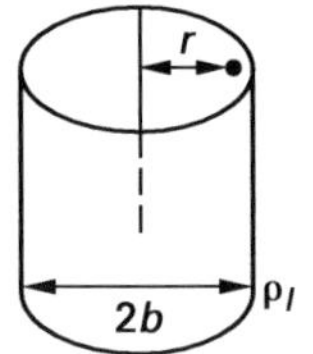 $E=\frac{\rho_l}{2\pi\epsilon r}$
concentric spheres 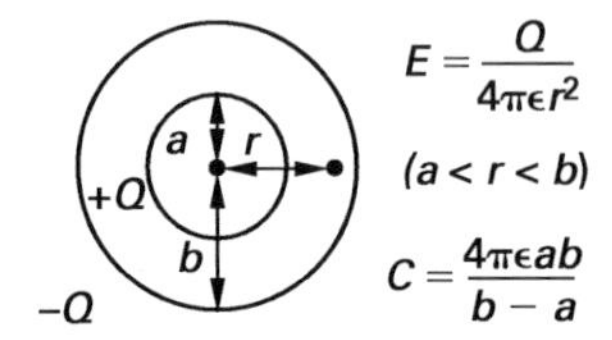 $E=\frac{Q}{4\pi\epsilon r^2}$ $(a<r<b)$ $C=\frac{4\pi\epsilon ab}{b-a}$	infinite sheet distribution 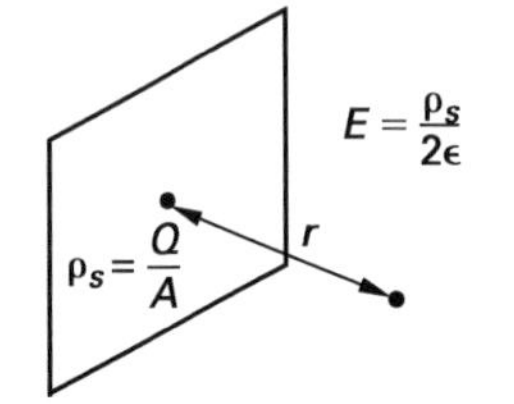 $E=\frac{\rho_s}{2\epsilon}$ $\rho_s=\frac{Q}{A}$
infinite line distribution 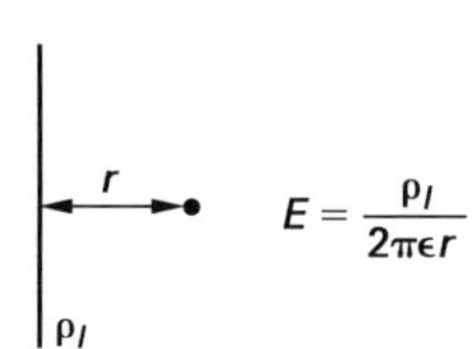 $E=\frac{\rho_l}{2\pi\epsilon r}$	infinite parallel plates 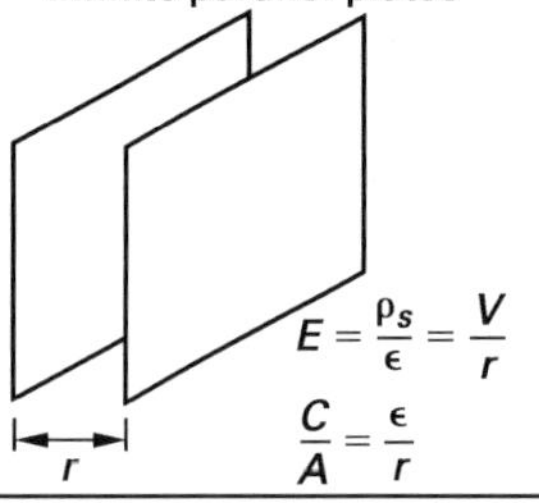 $E=\frac{\rho_s}{\epsilon}=\frac{V}{r}$ $\frac{C}{A}=\frac{\epsilon}{r}$
infinite isolated cylinder 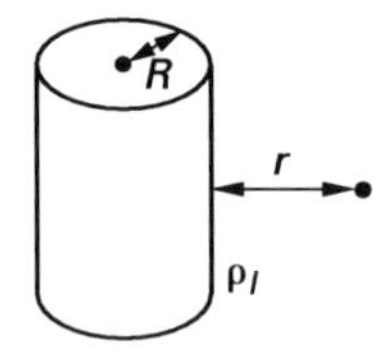 $E=\frac{\rho_l}{2\pi\epsilon(r+R)}\quad(r>0)$	dipole (doublet) 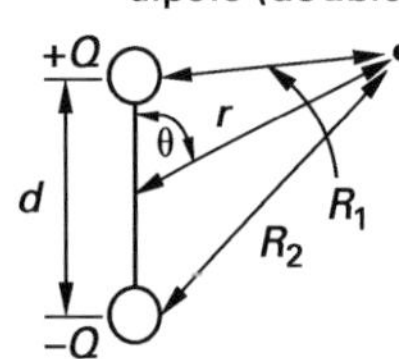 $E=\frac{Qd}{4\pi\epsilon r^3}(2\cos\theta a_r+\sin\theta a_\theta)$

For n capacitors in parallel,

$$C_{\text{total}}=C_1+C_2+C_3+\cdots+C_n \qquad 16.26$$

For n capacitors in series,

$$\frac{1}{C_{\text{total}}}=\frac{1}{C_1}+\frac{1}{C_2}+\frac{1}{C_3}+\cdots+\frac{1}{C_n} \qquad 16.27$$

15. ENERGY DENSITY IN AN ELECTRIC FIELD

$$\begin{aligned}U_{\text{ave}}&=\tfrac{1}{2}V_{\text{voltage}}Q=\tfrac{1}{2}(Et)(DA)=\tfrac{1}{2}EDV_{\text{volume}}\\&=\tfrac{1}{2}\epsilon E^2V_{\text{volume}}\\&=\tfrac{1}{2}\frac{D^2}{\epsilon}V_{\text{volume}}\end{aligned} \qquad 16.28$$

Equation 16.28 in all its forms assumes that **D** and **E** are in the same direction.

$$u_{\text{ave}}=\frac{U_{\text{ave}}}{V_{\text{volume}}}=\tfrac{1}{2}\epsilon E^2=\tfrac{1}{2}\frac{D^2}{\epsilon} \qquad 16.29$$

16. SPEED AND MOBILITY OF CHARGE CARRIERS

$$\mathbf{F}=Q\mathbf{E}=m\mathbf{a} \qquad 16.30$$

- *average drift velocity*

$$\begin{aligned}\mathbf{v}_d&=\int\mathbf{a}dt=\int\frac{\mathbf{F}}{m}dt=\frac{Q}{m}t_m\mathbf{E}\\&=\mu\mathbf{E}\end{aligned} \qquad 16.31$$

17. CURRENT

$$V=IR \qquad 16.32$$

For simple DC circuits,

$$I=\frac{dQ}{dt} \qquad 16.33$$

18. CONVECTION CURRENT

$$\begin{aligned}I&=\rho A\mathrm{v}_{\text{ave}}=\left(\frac{Q}{V}\right)A\mathrm{v}_{\text{ave}}=\left(\frac{Q}{Al}\right)A\left(\frac{l}{t}\right)\\&=\frac{Q}{t}\end{aligned} \qquad 16.34$$

In terms of the current density vector,

$$\mathbf{J}=\rho\mathbf{v}_d \qquad 16.35$$

Figure 16.9 *Convection Current*

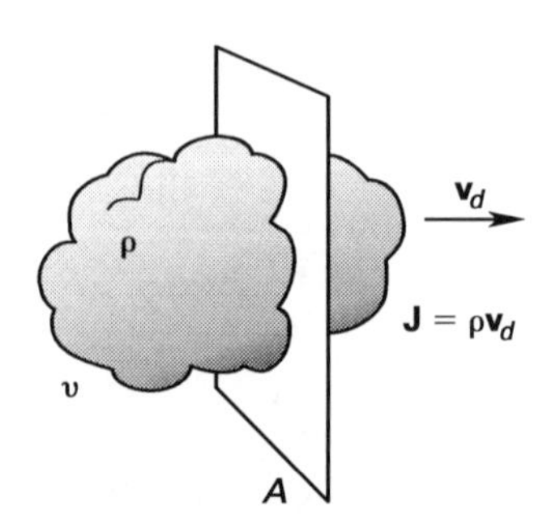

19. DISPLACEMENT CURRENT

$$i_d = \frac{d\psi}{dt} \quad 16.36$$

$$\mathbf{J}_d = \frac{\delta \mathbf{D}}{\delta t} \quad 16.37$$

$$i_d = \frac{d}{d_t}\int_A \mathbf{D}\cdot d\mathbf{A} \quad 16.38$$

20. CONDUCTION CURRENT

$$\mathbf{J}_c = \rho \mathbf{v}_d \quad 16.40$$

- *point form of Ohm's law*

$$\mathbf{J} = \sigma \mathbf{E} \quad 16.41$$

21. MAGNETIC POLES

$$m = pd \quad 16.42$$

22. BIOT-SAVART LAW

$$\mathbf{F}_{1\text{-}2} = \frac{\mu}{4\pi}\left(\frac{(I_2 d\mathbf{l}_2) \times (I_1 d\mathbf{l}_1) \times (\mathbf{r}_{1\text{-}2})}{r_{1\text{-}2}^2}\right) \quad 16.43$$

For parallel currents,

$$F_{1\text{-}2} = \frac{\mu}{4\pi r^2} I_1 dl_1 I_2 dl_2 \quad 16.44$$

- *magnetic force between two moving charges*

$$\mathbf{F}_{1\text{-}2} = \frac{\mu}{4\pi}\left(\frac{Q_1 Q_2}{r_{1-2}^2}\right)\mathbf{v}_2 \times (\mathbf{v}_1 \times \mathbf{r}_{1\text{-}2}) \quad 16.45$$

- *maximum magnetic force between parallel moving charges*

$$F = \frac{\mu}{4\pi r^2} Q_1 Q_2 \mathrm{v}^2 \quad 16.46$$

- *maximum coulombic force between moving charges in free space*

$$F_e = \frac{Q_1 Q_2}{4\pi\epsilon_0 r^2} \quad 16.47$$

- *maximum magnetic force between moving charges in free spaces*

$$F_m = \frac{\mu_0 Q_1 Q_2 \mathrm{v}^2}{4\pi r^2} \quad 16.48$$

$$\frac{F_m}{F_e} = \epsilon_0 \mu_0 \mathrm{v}^2 \quad 16.49$$

$$c^2 = \frac{1}{\mu_0 \epsilon_0} \quad 16.50$$

23. MAGNETIC FIELDS

$$\mathbf{B} = \frac{\mu I l}{4\pi r^2}\mathbf{a} \quad 16.52$$

$$B = |\mathbf{B}| = \frac{\phi}{A} \quad 16.53$$

24. PERMEABILITY AND SUSCEPTIBILITY

$$\mu = \mu_0 \mu_r \quad 16.55$$

$$\mu = \frac{B}{H} \quad 16.56$$

$$\mu = \mu_0(1 + \chi_m) \quad 16.57$$

- *magnetic susceptibility*, χ_m (a dimensionless quantity)

$$\chi = \frac{M}{H} \quad 16.58$$

25. MAGNETIC FLUX

The term NI is the *magnetomotive force.*

$$\phi = \mu N I l = BA \quad 16.60$$

26. MAGNETIC FIELD STRENGTH

(See Table 16.8.)

$$\mathbf{H} = \frac{1}{\mu}\mathbf{B} \quad 16.61$$

28. INDUCTANCE AND RECIPROCAL INDUCTANCE

$$\mathrm{v} = L\frac{di}{dt} \quad 16.63$$

29. INDUCTORS

For a solenoid,

$$L = \mu\frac{N^2 A_{\text{coil}}}{l} \quad 16.67$$

- *total energy*, U *(in joules)*

$$U = \tfrac{1}{2} L I^2 \quad 16.68$$

For inductors in series,

$$L_{\text{total}} = L_1 + L_2 + L_3 + \cdots + L_n \quad 16.69$$

For inductors in parallel,

$$\frac{1}{L_{\text{total}}} = \frac{1}{L_1} + \frac{1}{L_2} + \frac{1}{L_3} + \cdots + \frac{1}{L_n} \quad 16.70$$

Table 16.8 *Magnetic Field and Inductance for Various Configurations*

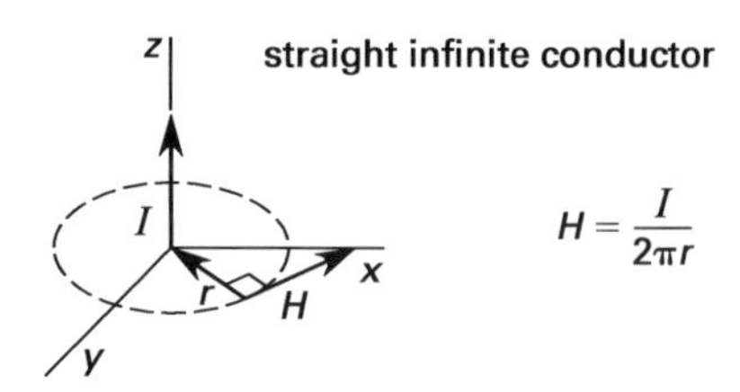

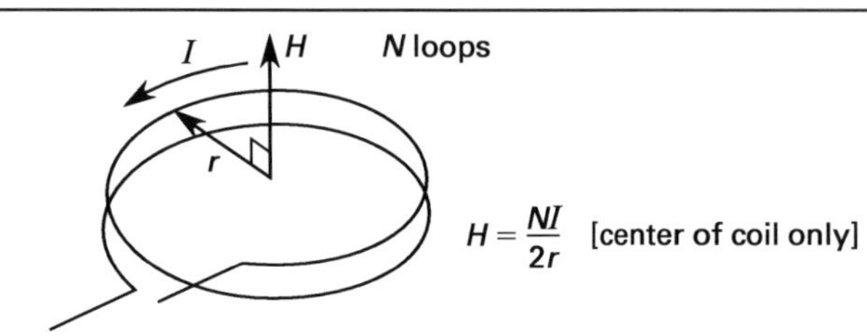

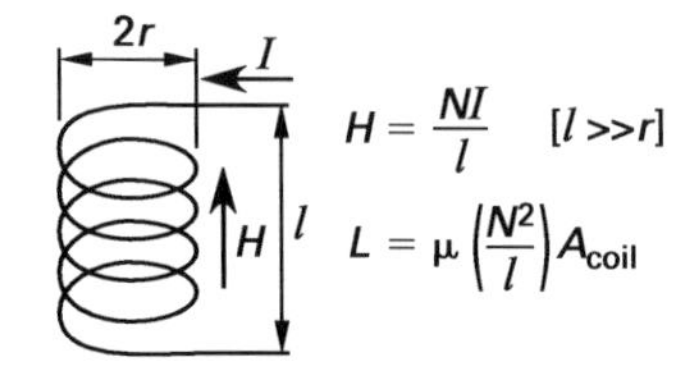

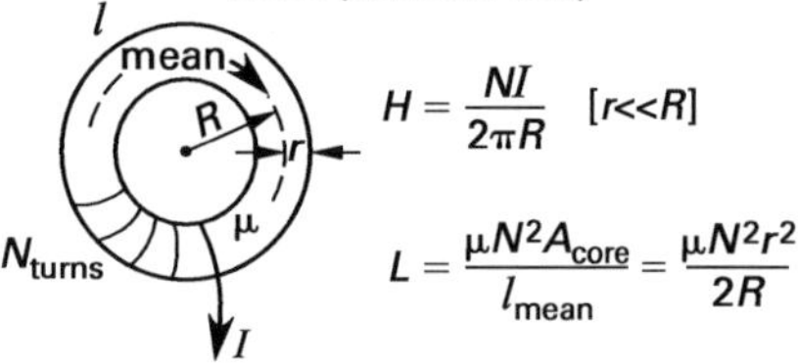

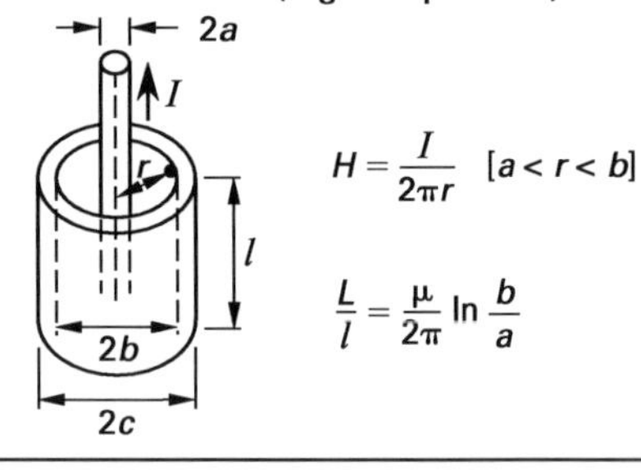

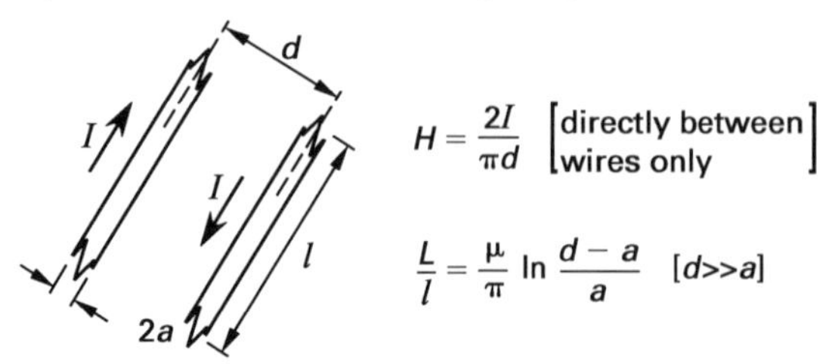

Reprinted with permission from *Core Engineering Concepts for Students and Professionals*, by Michael R. Lindeburg, PE, copyright 2010, by Professional Publications, Inc.

30. ENERGY DENSITY IN A MAGNETIC FIELD

$$U_{ave} = \tfrac{1}{2}\phi NI \qquad 16.71$$

$$u_{ave} = \frac{U_{ave}}{V_{volume}} = \tfrac{1}{2}BH = \tfrac{1}{2}\mu H^2 = \tfrac{1}{2}\frac{B^2}{\mu} \qquad 16.72$$

31. SPEED AND DIRECTION OF CHARGE CARRIERS

$$\mathbf{F} = Q\mathbf{v} \times \mathbf{B} \qquad 16.74$$

If the conductor is straight and the magnetic field is constant along the length of the conductor,

$$F = NIBl\sin\theta \qquad 16.76$$

If at right angles,

$$F = NIBl \qquad 16.77$$

- *Lorentz force equation*

$$\mathbf{F} = Q(\mathbf{E} + \mathbf{v} \times \mathbf{B}) \qquad 16.80$$

Electromagnetic waves propagate through space with a velocity given by

$$\mathrm{v} = \frac{1}{\sqrt{\epsilon\mu}} \qquad 16.81$$

32. VOLTAGE AND THE MAGNETIC CIRCUIT

Figure 16.17 *Magnetic-Electric Circuit Analogy*

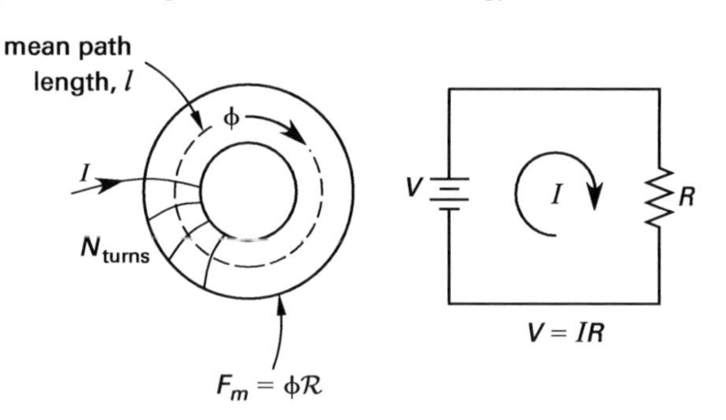

- *reluctance,* $\mathcal{R}$

$$\mathcal{R} = \frac{l}{\mu A} \qquad 16.84$$

- *magnetic equation that correlates with Ohm's law in electric circuits*

$$F_m = Hl = \phi\mathcal{R} \qquad 16.85$$

33. MAGNETIC FIELD-INDUCED VOLTAGE

$$V = -N\frac{d\phi}{dt} \qquad 16.86$$

$$V = N\frac{d\phi}{dt} = NBl\frac{ds}{dt} = NBl\mathrm{v} \qquad 16.87$$

EPRM Chapter 17
Electronic Theory

2. CHARGES IN A VACUUM

$$\mathbf{F}_{\parallel} = Q\mathbf{E} \quad 17.1$$

$$\mathbf{F}_{\perp} = Q\mathbf{v} \times \mathbf{B} \quad 17.2$$

The direction of motion of the particle will be a circle if the magnetic field is uniform and the particle enters it at right angles.

$$r = \frac{m\mathrm{v}}{QB} \quad 17.3$$

$$\mathbf{F} = \mathbf{F}_{\parallel} + \mathbf{F}_{\perp} = Q(\mathbf{E} + (\mathbf{v} \times \mathbf{B})) \quad 17.4$$

- *photoelectric effect*

$$h\nu = \varphi + \tfrac{1}{2}m\mathrm{v}^2 \quad 17.6$$

3. CHARGES IN LIQUIDS AND GASES

- *conductivity and resistivity*

$$\sigma = \rho\mu \quad 17.8$$

In metallic conductors, the conductivity is often defined by the point form of Ohm's law.

$$\sigma = \frac{|\mathbf{J}|}{|\mathbf{E}|} = \frac{J}{E} \quad 17.9$$

4. CHARGES IN SEMICONDUCTORS

- *law of mass action*

$$np = n_i^2 \quad 17.13$$

- *Einstein equation*

$$V_T = \frac{D_p}{\mu_p} = \frac{D_n}{\mu_n} \quad 17.15$$

$$V_T = \frac{kT}{q} \quad 17.16$$

EPRM Chapter 18
Communication Theory

1. FUNDAMENTALS

Analog and Digital Signals

- *Nyquist rate*, f_S (rate that achieves this ideal sampling)

$$f_S = \frac{1}{T_S} = 2f_I \quad 18.1$$

- *Nyquist interval*, T_S (the sampling period corresponding to the Nyquist rate)

$$T_S = \frac{1}{2f_I} \quad 18.2$$

Basic Signal Theory

- *primary uncertainty relationship*

$$\Delta t \Delta\Omega = 2\pi \quad 18.4$$

- *second moment uncertainty relationship* (which relates the energy)

$$\Delta t_2 \Delta\Omega_2 \geq \frac{1}{2} \quad 18.5$$

Information Entropy

- *probability of occurrence*, $p(x_i)$
- *self-information*, $I(x_i)$

$$I(x_i) = \log\frac{1}{p(x_i)} = -\log p(x_i) \quad 18.6$$

Decibels

$$r = 10\log_{10}\frac{P_2}{P_1} \quad 18.8$$

$$r = 10\log\left(\frac{I_2}{I_1}\right)^2 = 10\log\left(\frac{V_2}{V_1}\right)^2 = 20\log\frac{I_2}{I_1}$$

$$= 20\log\frac{V_2}{V_1} \quad 18.9$$

4. NOISE

- *thermal noise* (or *Johnson noise*)

$$\overline{v_n^2} = 4kTR(\mathrm{BW}) \quad 18.10$$

- *shot noise* (or *Schottky noise*)

$$I = q\frac{dn}{dt} \quad 18.11$$

EPRM Chapter 19
Acoustic and Transducer Theory

1. INTRODUCTION

$$\mathrm{v}_s = f\lambda \quad 19.1$$

$$T = \frac{1}{f} \quad 19.2$$

2. SOUND WAVE PROPAGATION VELOCITY

- *speed of sound in solids and liquids*

$$\mathrm{v}_s = \sqrt{\frac{E}{\rho}} \quad 19.3$$

- *propagation velocity* (for ideal gases)

$$\mathrm{v}_s = \sqrt{\frac{\gamma R^* T}{\mathrm{MW}}} = \sqrt{\gamma R T} \qquad 19.4$$

$$\gamma = \frac{c_p}{c_v}$$

3. ENERGY AND INTENSITY OF SOUND WAVES

- *energy density*, η

$$\eta = \tfrac{1}{2}\frac{p_0^2}{\rho \mathrm{v}^2} \qquad 19.5$$

- *intensity of the sound wave*

$$I = \eta \mathrm{v} = \tfrac{1}{2}\frac{p_0^2}{\rho \mathrm{v}} \qquad 19.6$$

- *sound-pressure level*, L_p

$$L_p = 20 \log \frac{p}{p_0} \qquad 19.8$$

Field Theory

EPRM Chapter 20
Electrostatics

PART 1. ELECTROSTATIC FIELDS

$$\mathbf{E}_1 = \frac{\mathbf{F}_{1-2}}{\Delta Q_2} = \left(\frac{Q_1}{4\pi\epsilon r^2}\right)\mathbf{a} \qquad 20.1$$

$$\mathbf{F} = Q\mathbf{E} \qquad 20.2$$

Figure 20.2 *Electric Field Interaction*

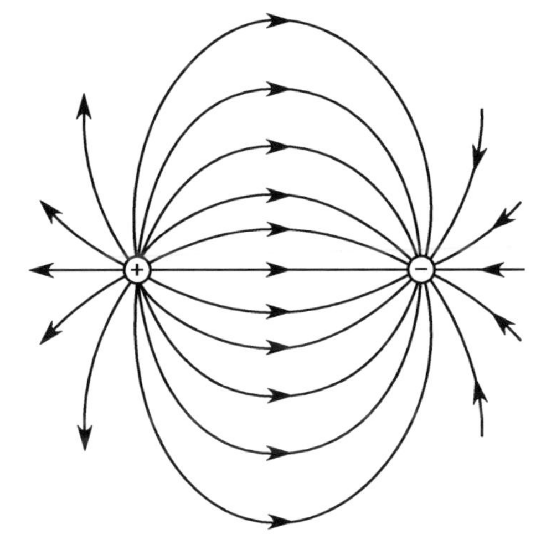

(a) unlike charges

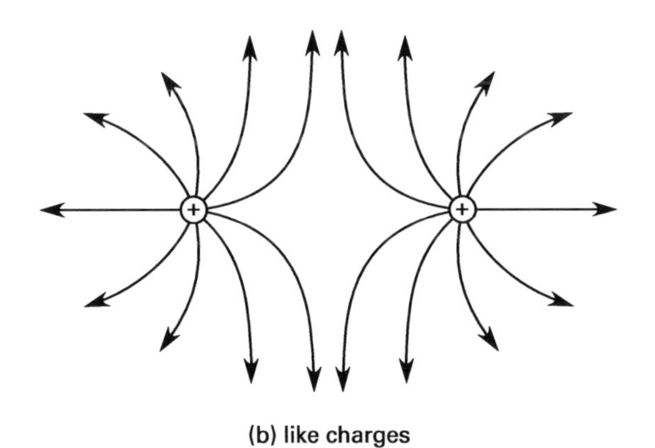

(b) like charges

1. POINT CHARGE

$$\mathbf{E} = \left(\frac{Q}{4\pi\epsilon r^2}\right)\mathbf{a}_r \qquad 20.4$$

$$V = \frac{Q}{4\pi\epsilon r} \qquad 20.5$$

2. DIPOLE CHARGE

$$\mathbf{p} = Q\mathbf{d} \qquad 20.6$$

$$\mathbf{E} = \left(\frac{Qd}{4\pi\epsilon r^3}\right)(2\cos\theta\mathbf{a}_r + \sin\theta\mathbf{a}_\theta) \qquad 20.7$$

$$V = \frac{Q(R_2 - R_1)}{4\pi\epsilon R_1 R_2} \approx \frac{Qd\cos\theta}{4\pi\epsilon r^2} \qquad 20.8$$

3. COULOMB'S LAW

$$\mathbf{F} = \left(\frac{Q_1 Q_2}{4\pi\epsilon r^2}\right)\mathbf{a} \qquad 20.11$$

4. LINE CHARGE

$$\mathbf{E} = \left(\frac{\rho_l}{2\pi\epsilon r}\right)\mathbf{a}_r \qquad 20.12$$

5. PLANE CHARGE

$$\mathbf{E} = \left(\frac{\rho_s}{2\epsilon}\right)\mathbf{a}_n \qquad 20.14$$

PART 2. DIVERGENCE

$$\text{div } \mathbf{D} = \nabla\cdot\mathbf{D} = \rho \qquad 20.18$$

PART 3. POTENTIAL

- *potential* (in terms of the electric field)

$$\mathbf{E} = -\nabla V \qquad 20.19$$

- *potential* (in integral form)

$$V_{\text{BA}} = \frac{W}{\Delta Q} = -\int_{\text{B}}^{\text{A}} \mathbf{E}\cdot d\mathbf{l} \qquad 20.20$$

$$\text{div } \mathbf{E} = \nabla\cdot\mathbf{E} = \frac{\rho}{\epsilon} \qquad 20.21$$

$$\nabla^2 v = -\frac{\rho}{\epsilon} \qquad 20.22$$

PART 4. WORK AND ENERGY

- *total work done in moving a charge from point A to point B in an electric field*

$$W_{AB} = -\int_A^B Q\mathbf{E}\cdot d\mathbf{l} = -\int_A^B QE\cos\theta dl \qquad 20.25$$

- *energy "stored" in the volume of the electrostatic field*

$$U = \tfrac{1}{2}\int \rho V dv = \tfrac{1}{2}\int \mathbf{D}\cdot\mathbf{E} dv = \tfrac{1}{2}\int \epsilon E^2 dv = \tfrac{1}{2}\int \frac{D^2}{\epsilon} dv \qquad 20.26$$

$$W = -\oint Q\mathbf{E}\cdot d\mathbf{l} = 0 \qquad 20.27$$

EPRM Chapter 21
Electrostatic Fields

1. POLARIZATION

$$\mathbf{P} = \chi_e \epsilon_0 \mathbf{E} \qquad 21.3$$

$$\mathbf{D} = \epsilon_0(1+\chi_e)\mathbf{E} \qquad 21.4$$

$$\mathbf{D} = \epsilon\mathbf{E} = \epsilon_0\epsilon_r\mathbf{E} = \epsilon_0\mathbf{E} + \mathbf{P} \qquad 21.7$$

3. ELECTRIC DISPLACEMENT

- *work of rotating a dipole* (if the angle $\pi/2$ is chosen as the arbitrary zero energy reference)

$$W = -\mathbf{p}\cdot\mathbf{E} \qquad 21.8$$

For a poor conductor, or *lossy dielectric*,

$$\frac{J_c}{J_d} = \frac{\sigma}{\omega\epsilon} = \frac{\sigma}{2\pi f\epsilon_0\epsilon_r} \qquad 21.10$$

EPRM Chapter 22
Magnetostatics

PART 1. THE MAGNETIC FIELD

The strength of the **B**-field is

$$\mathbf{B} = \left(\frac{\mu I l}{4\pi r^2}\right)\mathbf{a} \qquad 22.1$$

The term **B** is most often called the magnetic flux density.

$$B = \frac{\Phi}{A} \qquad 22.2$$

1. BIOT-SAVART LAW

$$\mathbf{B} = \frac{\mu_0}{4\pi}\left(\frac{q\mathbf{v}\times\mathbf{r}}{r^2}\right) \qquad 22.7$$

2. FORCE ON A MOVING CHARGED PARTICLE

$$\mathbf{F} = Q\mathbf{v}\times\mathbf{B} \qquad 22.8$$

- *radius and an angular velocity of a charged particle in a uniform magnetic field traveling in a circular path*

$$r = \frac{m\mathrm{v}}{QB} \qquad 22.9$$

$$\omega = \frac{QB}{m} \qquad 22.10$$

$$B = \frac{F}{q\mathrm{v}} \qquad 22.11$$

3. FORCE ON CURRENT ELEMENTS

(See Fig. 22.2.)

In terms of magnitude, assuming a uniform magnetic field,

$$F = lIB\sin\theta \qquad 22.15$$

For a straight, infinitely long conductor,

$$B = \frac{\mu I}{2\pi r} \qquad 22.16$$

For two straight, infinitely long conductors,

$$\frac{\mathbf{F}}{l} = \mathbf{I}\times\mathbf{B} = \left(\frac{\mu I_1 I_2}{2\pi r}\right)\mathbf{a} \qquad 22.17$$

PART 2. LORENTZ FORCE LAW

$$\mathbf{F} = \mathbf{F}_e + \mathbf{F}_m = Q(\mathbf{E} + \mathbf{v}\times\mathbf{B}) \qquad 22.19$$

$$d\mathbf{F} = \rho(\mathbf{E} + \mathbf{v}\times\mathbf{B})dV \qquad 22.20$$

PART 3. TRADITIONAL MAGNETISM

(See Table 22.1.)

$$\mathbf{T} = q_m\mathbf{d}\times\mathbf{B} \qquad 22.22$$

$$B = \frac{pl}{lA} = \frac{p}{A} = \frac{\mu m}{V} \qquad 22.23$$

6. MAGNETIC DIPOLE

- *magnetic moment*, **m**

$$\mathbf{m} = IA\mathbf{n} \qquad 22.27$$

$$\mathbf{T} = \mathbf{m}\times\mathbf{B} \qquad 22.28$$

Figure 22.2 Magnetic Force Direction

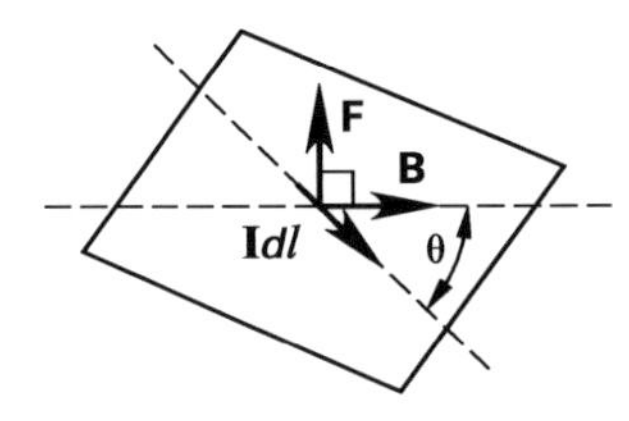

(a) magnetic force direction in terms of I*dl*

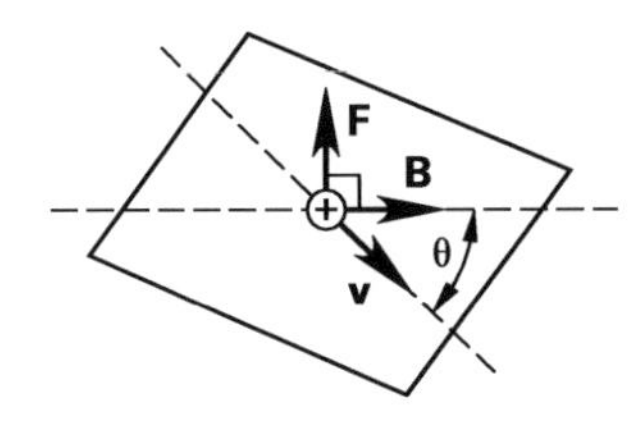

(b) magnetic force direction in terms of individual charge velocities

When the left hand index finger points in the direction of **B**, and the third finger points along **I**, the thumb points in the **F** direction.

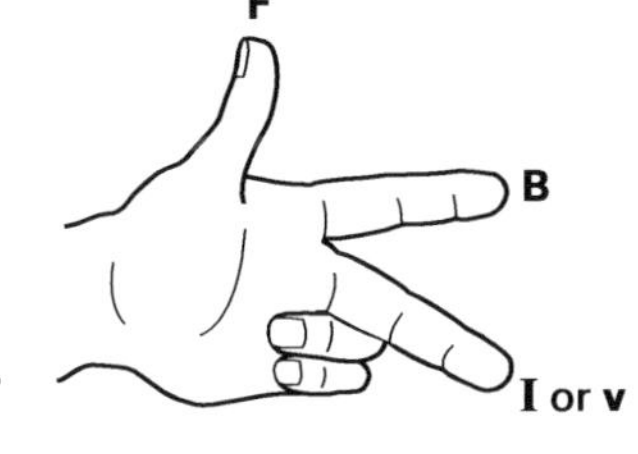

(c) FBI rule

Rotate the **I** vector into the **B** vector, and position your hand so that your fingers curl in the same direction as the **I** vector rotates. Your extended thumb will coincide with the direction of moment.

F

B

I or v

(d) right-hand rule

7. COULOMB'S LAW EQUIVALENT: FORCE BETWEEN MAGNETIC POLES

$$\mathbf{F}_{1-2} = p_2\mathbf{H}_1 = \left(\frac{p_1 p_2}{4\pi\mu r^2}\right)\mathbf{a} \qquad 22.30$$

PART 4. CURL

In magnetostatics, no magnetic charge exists.

$$\nabla\cdot\mathbf{B} = 0 \qquad 22.38$$

- *curl of the* **B**-*field in free space*

$$\nabla\times\mathbf{B} = \mu_0\mathbf{J} \qquad 22.39$$

In terms of the magnetic field strength,

$$\nabla\times\mathbf{H} = \mathbf{J} \qquad 22.40$$

EPRM Chapter 23 Magnetostatic Fields

1. MAGNETIZATION

$$\mathbf{B} = \mu\mathbf{H} = \mu_0\mu_r\mathbf{H} = \mu_0\mathbf{H} + \mu_0\mathbf{M} \qquad 23.8$$

3. AUXILIARY FIELD H

$$H = \frac{NI}{l} \qquad 23.10$$

EPRM Chapter 24 Electrodynamics

1. ELECTROMOTIVE FORCE

$$\begin{aligned}\mathcal{E} &= \oint \mathbf{F}\cdot d\mathbf{l} = \oint \frac{\mathbf{J}}{\sigma}\cdot d\mathbf{l} \\ &= \oint \frac{I}{A\sigma}dl = I\oint \frac{1}{A\sigma}dl = IR \end{aligned} \qquad 24.1$$

Electromotive force can be generated by time-varying flux.

$$v = -N\frac{d\phi}{dt} \qquad 24.2$$

$$v = -N\frac{d\phi}{di}\frac{di}{dt} = -L\frac{di}{dt} \qquad 24.3$$

Electromagnetic force can be generated by the flux-cutting method.

$$V = -NBl\mathrm{v} \qquad 24.4$$

3. FARADAY'S LAW OF ELECTROLYSIS

- *Faraday's law of electrolysis*

$$m_g = \frac{ItW_e}{F} \qquad 24.8$$

The term W_e is the equivalent weight of an element, the atomic weight, AW, in grams divided by the valence change that occurs during electrolysis, z.

$$W_e = \frac{\mathrm{AW}}{z} \qquad 24.9$$

Table 22.1 Magnetic Units

quantity	symbol	cgs units	SI units	conversion: cgs to SI
pole strength or flux	p or Φ or ϕ[a]	maxwells[b]	Wb	10^8 maxwells = 1 Wb
flux density	B	gauss[c]	T[d]	10^4 gauss = 1 T
field strength (intensity)	H	oersted	A/m[e]	$4\pi \times 10^{-3}$ oersted = 1 A/m
magnetization[f]	M	oersted	A/m[e]	$4\pi \times 10^{-3}$ oersted = 1 A/m
permeability	μ	gauss/oersted	H/m[g]	$\mu_{cgs}\,(4\pi \times 10^{-7}) = \mu_{mks}$[h]

[a]The SI symbol for magnetic flux is Φ. The symbol p is used when referring to magnetic poles. The symbol ϕ is a generic flux symbol used in many texts.
[b]A *maxwell* is a line of force.
[c]A gauss is equivalent to lines/cm^2 or maxwells/cm^2.
[d]A tesla is equivalent to Wb/m^2.
[e]A/m is equivalent to N/Wb.
[f]The units used in this row assume the following equation for flux density: $B = \mu_0 H + \mu_0 M$. Other forms of this equation exist. Use care in determining the correct units or in comparing values from different references.
[g]H/m is equivalent to Wb/A·m and Ω·s/m.
[h]The value of μ_{cgs} in the cgs system is 1.

Table 25.1 Maxwell's Equations

integral form	point form	remarks
$\oint_s \mathbf{D}\cdot d\mathbf{s} = \int_V \rho\, dv$	$\nabla\cdot\mathbf{D} = \rho$	Gauss' law
$\oint_s \mathbf{B}\cdot d\mathbf{s} = 0$	$\nabla\cdot\mathbf{B} = 0$	nonexistence of magnetic monopoles
$\oint \mathbf{E}\cdot d\mathbf{l} = \int_s \left(\frac{-\partial\mathbf{B}}{\partial t}\right)\cdot d\mathbf{s}$	$\nabla\times\mathbf{E} = -\frac{\partial\mathbf{B}}{\partial t}$	Faraday's law
$\oint \mathbf{H}\cdot d\mathbf{l} = \int_s \left(\mathbf{J}_c + \frac{\partial\mathbf{D}}{\partial t}\right)\cdot d\mathbf{s}$	$\nabla\times\mathbf{H} = \mathbf{J}_c + \frac{\partial\mathbf{D}}{dt}$	Ampère's law

5. ENERGY AND MOMENTUM: POYNTING'S VECTOR

$$\mathbf{S} = c\epsilon E^2\mathbf{a} = c\mu H^2\mathbf{a} = \mathbf{E}\times\mathbf{H} \qquad 24.12$$

$$\mathcal{P}_{\text{ave}} = \tfrac{1}{2}\text{Re}(\mathbf{E}\times\mathbf{H}^*) \qquad 24.13$$

EPRM Chapter 25 Maxwell's Equations

1. MAXWELL'S EQUATIONS

(See Table 25.1.)

Table 25.2 Maxwell's Equations: Free-Space Form

integral form	point form
$\oint_s \mathbf{D}\cdot d\mathbf{s} = 0$	$\nabla\cdot\mathbf{D} = 0$
$\oint_s \mathbf{B}\cdot d\mathbf{s} = 0$	$\nabla\cdot\mathbf{B} = 0$
$\oint \mathbf{E}\cdot d\mathbf{l} = \int_s \left(\frac{-\partial\mathbf{B}}{\partial t}\right)\cdot d\mathbf{s}$	$\nabla\times\mathbf{E} = -\frac{\partial\mathbf{B}}{\partial t}$
$\oint \mathbf{H}\cdot d\mathbf{l} = \int_s \left(\frac{\partial\mathbf{D}}{\partial t}\right)\cdot d\mathbf{s}$	$\nabla\times\mathbf{H} = \frac{\partial\mathbf{D}}{dt}$

2. ELECTROMAGNETIC FIELD VECTORS

Table 25.3 Electromagnetic Field Vector Equations

$$\mathbf{D} = \epsilon\mathbf{E} = \epsilon_0\mathbf{E} + \mathbf{P} = \epsilon_0(1+\chi_e)\mathbf{E}$$

$$\mathbf{B} = \mu\mathbf{H} = \mu_0\mathbf{H} + \mu_0\mathbf{M} = \mu_0(1+\chi_m)\mathbf{H}$$

$$\mathbf{J} = \sigma\mathbf{E} = \rho\mathrm{v}$$

3. COMPARISON OF ELECTRIC AND MAGNETIC EQUATIONS

Table 25.4 Electric and Magnetic Circuit Analogies

electric	magnetic
emf $= V = IR$	mmf $= V_m = \phi\mathcal{R}$
current I	flux ϕ
emf $\mathcal{E}$ or V	mmf V_m
resistance $R = \rho l/A = l/\sigma A$	reluctance $\mathcal{R} = l/\mu A$
resistivity ρ	reluctivity $1/\mu$
conductance $G = 1/R$	permeance $P_m = \mu A/l$
conductivity $\sigma = 1/\rho$	permeability μ

(See the "Comparison of Electric and Magnetic Equations" table.)

Comparison of Electric and Magnetic Equations

equation description	electric version	magnetic version	remarks
experimental force law	Coulomb's law $\mathbf{F} = \left(\frac{Q_1 Q_2}{4\pi r^2}\right)\mathbf{r}$ 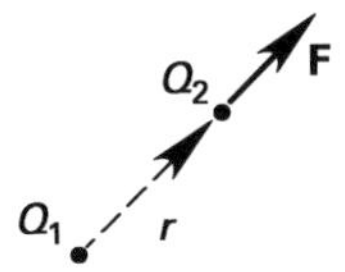	force between two current elements $d\mathbf{F} = \left(\frac{\mu_0}{4\pi}\right)\left(\frac{I_2 d\mathbf{l}_2 \times (I_1 d\mathbf{l}_1 \times r)}{r^2}\right)$ 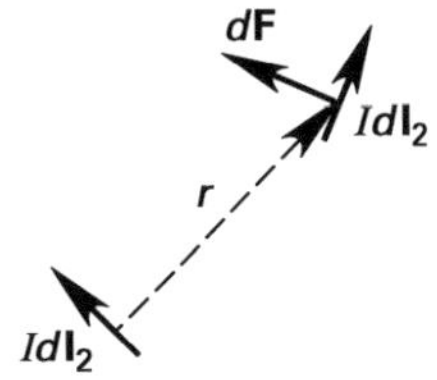	The term $Id\mathbf{l}$ in the magnetic column is the equivalent of a "magnetic charge" q_m. The I or the $d\mathbf{l}$ can be the vector. The $\mathbf{r}$ is a unit vector pointing from 1 to 2.
field definitions from force law	$\mathbf{F} = Q\mathbf{E}$	$d\mathbf{F} = \mathbf{I} \times \mathbf{B}d\mathbf{l}$ current element $d\mathbf{F} = \mathbf{J} \times \mathbf{B}dV$ distributed current element $d\mathbf{F} = q\mathbf{v} \times \mathbf{B}$ moving charge	The V used in this row represents volume, not voltage. The $\mathbf{v}$ is the velocity.
general force law	$\mathbf{F} = q(\mathbf{E} + \mathbf{v} \times \mathbf{B})$ $d\mathbf{F} = (\rho\mathbf{E} + \mathbf{J} \times \mathbf{B})dV$ where $dQ = \rho dV$		The V in this row represents the volume, not voltage. The $\mathbf{v}$ is the velocity.
definition of scalar and vector potential	$E = -\nabla V$	$\mathbf{B} = \nabla \times \mathbf{A}$	$\mathbf{A}$ is the magnetic vector potential. V in this row represents voltage.
Poisson's equation for the potential function	$\nabla^2 V = -\frac{\rho}{\epsilon}$	$\nabla^2 \mathbf{A} = -\mu_0 \mathbf{J}$	From a knowledge of the charge distribution, the potential can be found and then the $\mathbf{E}$ and $\mathbf{B}$ fields determined.
Gauss's law enclosing charge and Ampere's law enclosing current	$\iint \mathbf{D} \cdot d\mathbf{A} = \iiint \rho dV = Q$ $\nabla \cdot \mathbf{D} = \rho$	$\oint H \cdot d\mathbf{l} = I$ $\nabla \times \mathbf{H} = \mathbf{J}$	The V in this row represents volume.
constitutive relations	$\mathbf{D} = \epsilon\mathbf{E}$ $\mathbf{D} = \epsilon_0\mathbf{E} + \mathbf{P}$	$\mathbf{B} = \mu\mathbf{H}$ $\mathbf{B} = \mu_0\mathbf{H} + \mu_0\mathbf{M}$	The second set of equations is always valid. The first set assumes the medium is linear and isotropic.
definitions of relative permittivity and permeability	$\epsilon_r = \frac{\epsilon}{\epsilon_0}$ $\epsilon_0 = 8.854 \times 10^{-12}$ F/m	$\mu_r = \frac{\mu}{\mu_0}$ $\mu_0 = 4\pi \times 10^{-7}$ H/m	

(continued)

equation description	electric version	magnetic version	remarks
capacitance and inductance of a field cell	$\epsilon_0 = \frac{C}{l}$	$\mu_0 = \frac{L}{l}$	Field cells are a construct designed to represent free space in terms of a parallel plate capacitor and an inductor. This capacitance and inductance exist regardless of the presence of an electric or magnetic field.
capacitance and inductance	$C = \frac{Q}{V}$	$L = \frac{\Lambda}{I}$	Λ is the flux linkage.
energy density of a field	$U = \frac{1}{2}\epsilon E^2$	$U = \frac{1}{2}\mu H^2$	Both energy and momentum are carried by a field.
energy stored by capacitance and inductance	$W = \frac{1}{2}CV^2$	$W = \frac{1}{2}LI^2$	
electromotive and magnetomotive force with sources present	$\oint \mathcal{E} \cdot d\mathbf{l} = \mathcal{E} = V$	$\oint \mathbf{H} \cdot d\mathbf{l} = NI = F_m = V_m$	The $\mathcal{E}$ is the emf, not the permittivity. Without sources present, both line integrals are equal to zero.
dipole moments	$\mathbf{p} = q\mathbf{d}$ d	$\mathbf{m} = I\mathbf{A}$ m A I	
dipole torque	$\mathbf{T} = \mathbf{p} \times \mathbf{E}$	$\mathbf{T} = \mathbf{m} \times \mathbf{B}$	This torque occurs due to the dipole being immersed in an external **E** or **B** field.
dipole potential energy	$W = -\mathbf{p} \cdot \mathbf{E}$	$W = -\mathbf{m} \cdot \mathbf{B}$	

Circuit Theory

EPRM Chapter 26
DC Circuit Fundamentals

3. RESISTANCE

$$R = \frac{\rho l}{A} \quad 26.1$$

$$A_{\text{cmil}} = \left(\frac{d_{\text{inches}}}{0.001}\right)^2 \quad 26.2$$

- *thermal coefficient of resistivity, α*

$$R = R_0(1 + \alpha\Delta T) \quad 26.5$$

$$\rho = \rho_0(1 + \alpha\Delta T) \quad 26.6$$

4. CONDUCTANCE

$$\sigma = \frac{1}{\rho} \quad 26.7$$

$$\begin{aligned}\%\ \text{conductivity} &= \frac{\sigma}{\sigma_{\text{Cu}}} \times 100\% \\ &= \frac{\rho_{\text{Cu}}}{\rho} \times 100\% \end{aligned} \quad 26.9$$

$$\begin{aligned}\rho_{\text{Cu},20^\circ\text{C}} &= 1.7241 \times 10^{-6}\ \Omega\cdot\text{cm} \\ &= 0.3403\ \Omega\cdot\text{cmil/cm}\end{aligned} \quad 26.10$$

5. OHM'S LAW

$$V = IR \quad 26.11$$

6. POWER

$$\begin{aligned}P &= \frac{W}{t} = \frac{VIt}{t} \\ &= VI = \frac{V^2}{R}\end{aligned} \quad 26.12$$

$$P = I^2R \quad 26.13$$

8. ENERGY SOURCES

$$\text{VR} = \frac{V_{\text{nl}} - V_{\text{fl}}}{V_{\text{fl}}} \times 100\% \quad 26.15$$

13. VOLTAGE AND CURRENT DIVIDERS

$$V_2 = V_s\left(\frac{R_2}{R_1 + R_2}\right) \quad 26.21$$

Figure 26.5 *Divider Circuits*

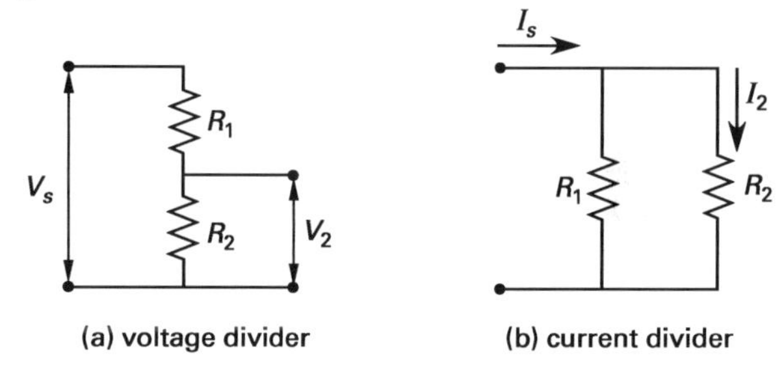

$$I_2 = I_s\left(\frac{R_1}{R_1 + R_2}\right) = I_s\left(\frac{G_2}{G_1 + G_2}\right) \quad 26.23$$

15. KIRCHHOFF'S VOLTAGE LAW

$$\sum_{\text{loop}} \text{voltage rises} = \sum_{\text{loop}} \text{voltage drops} \quad 26.24$$

16. KIRCHHOFF'S CURRENT LAW

$$\sum_{\text{node}} \text{currents in} = \sum_{\text{node}} \text{currents out} \quad 26.25$$

17. SERIES CIRCUITS

$$I = I_{R_1} = I_{R_2} = I_{R_3} \cdots = I_{R_n} \quad 26.26$$

$$R_e = R_1 + R_2 + R_3 + \cdots + R_n \quad 26.27$$

$$V_e = \pm V_1 \pm V_2 \pm \cdots \pm V_n \quad 26.28$$

$$V_e = IR_e \quad 26.29$$

18. PARALLEL CIRCUITS

$$V = V_1 = V_2 = V_3 \cdots = V_n$$
$$= I_1R_1 = I_2R_2 = I_3R_3 \cdots = V_n \qquad 26.30$$

$$\frac{1}{R_e} = \frac{1}{R_1} + \frac{1}{R_2} + \frac{1}{R_3} + \cdots + \frac{1}{R_n} \qquad 26.31(a)$$

$$G_e = G_1 + G_2 + G_3 + \cdots + G_n \qquad 26.31(b)$$

$$I = I_1 + I_2 + I_3 + \cdots + I_n$$
$$= \frac{V}{R_1} + \frac{V}{R_2} + \frac{V}{R_3} + \cdots + \frac{V}{R_n}$$
$$= V(G_1 + G_2 + G_3 + \cdots + G_n) \qquad 26.32$$

Figure 26.10 *Delta (Pi)-Wye (T) Configurations*

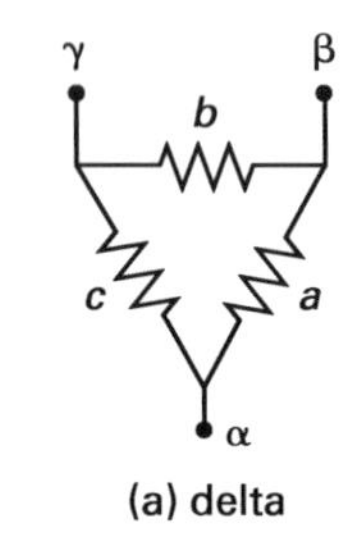

(a) delta

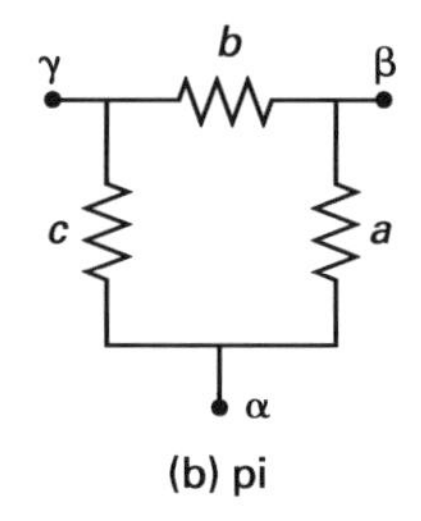

(b) pi

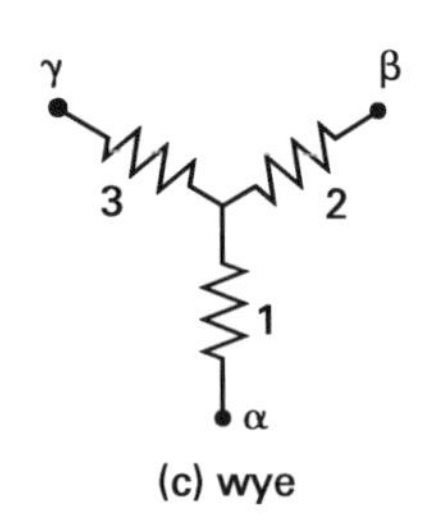

(c) wye

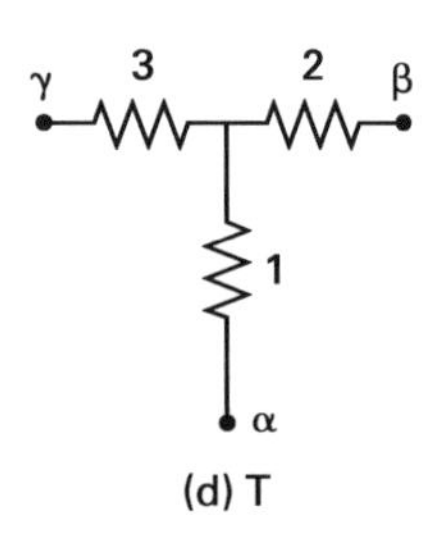

(d) T

20. DELTA-WYE TRANSFORMATIONS

$$R_a = \frac{R_1R_2 + R_1R_3 + R_2R_3}{R_3} \qquad 26.33$$

$$R_b = \frac{R_1R_2 + R_1R_3 + R_2R_3}{R_1} \qquad 26.34$$

$$R_c = \frac{R_1R_2 + R_1R_3 + R_2R_3}{R_2} \qquad 26.35$$

$$R_1 = \frac{R_aR_c}{R_a + R_b + R_c} \qquad 26.36$$

$$R_2 = \frac{R_aR_b}{R_a + R_b + R_c} \qquad 26.37$$

$$R_3 = \frac{R_bR_c}{R_a + R_b + R_c} \qquad 26.38$$

EPRM Chapter 27
AC Circuit Fundamentals

2. VOLTAGE

$$\omega = 2\pi f = \frac{2\pi}{T} \qquad 27.3$$

- *trigonometric:* $V_m \sin(\omega t + \theta)$
- *exponential:* $V_m e^{j\theta}$
- *polar* or *phasor:* $V_m \angle \theta$
- *rectangular:* $V_r + jV_i$

4. IMPEDANCE

$$\mathbf{Z} \equiv R + jX \qquad 27.6$$

$$R = Z\cos\phi \quad \left[\begin{array}{c}\text{resistive or}\\ \text{real part}\end{array}\right] \qquad 27.7$$

$$X = Z\sin\phi \quad \left[\begin{array}{c}\text{reactive or}\\ \text{imaginary part}\end{array}\right] \qquad 27.8$$

Table 27.1 *Characteristics of Resistors, Capacitors, and Inductors*

	resistor	capacitor	inductor
value	R (Ω)	C (F)	L (H)
reactance, X	0	$\frac{-1}{\omega C}$	ωL
rectangular impedance, $\mathbf{Z}$	$R + j0$	$0 - \frac{j}{\omega C}$	$0 + j\omega L$
phasor impedance, $\mathbf{Z}$	$R\angle 0°$	$\frac{1}{\omega C}\angle -90°$	$\omega L\angle 90°$
phase	in-phase	leading	lagging
rectangular admittance, $\mathbf{Y}$	$\frac{1}{R} + j0$	$0 + j\omega C$	$0 - \frac{j}{\omega L}$
phasor admittance, $\mathbf{Y}$	$\frac{1}{R}\angle 0°$	$\omega C\angle 90°$	$\frac{1}{\omega L}\angle -90°$

7. AVERAGE VALUE

$$f_{\text{ave}} = \frac{1}{T}\int_{t_1}^{t_1+T} f(t)\,dt \qquad 27.14$$

$$f_{\text{ave}} = \frac{\text{positive area} - \text{negative area}}{T} \qquad 27.15$$

$$V_{\text{ave}} = \frac{1}{\pi}\int_0^{\pi} v(\theta)\,d\theta = \frac{2V_m}{\pi} \quad \text{[rectified sinusoid]} \qquad 27.17$$

8. ROOT-MEAN-SQUARE VALUE

$$f_{\text{rms}}^2 = \frac{1}{T}\int_{t_1}^{t_1+T} f^2(t)\,dt \qquad 27.18$$

- *form factor*

$$\text{FF} = \frac{V_{\text{eff}}}{V_{\text{ave}}} \qquad 27.20$$

- *crest factor* CF (also known as the *peak factor* or *amplitude factor*)

$$\text{CF} = \frac{V_m}{V_{\text{eff}}} \qquad 27.21$$

(See Table 27.2.)

9. PHASE ANGLES

A leading circuit is termed a *capacitive circuit.* A lagging circuit is termed an *inductive circuit.*

$$v(t) = V_m \sin(\omega t + \theta) \quad \text{[reference]} \qquad 27.24$$

$$i(t) = I_m \sin(\omega t + \theta + \phi) \quad \text{[leading]} \qquad 27.25$$

$$i(t) = I_m \sin(\omega t + \theta - \phi) \quad \text{[lagging]} \qquad 27.26$$

12. COMPLEX REPRESENTATION

(See Table 27.3.)

Figure 27.9 *Complex Quantities*

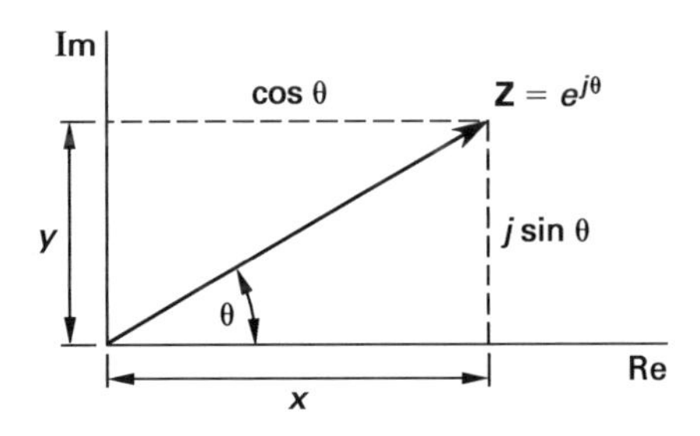

17. OHM'S LAW

$$\mathbf{V} = \mathbf{IZ} \qquad 27.41$$

Table 27.2 *Characteristics of Alternating Waveforms*

waveform	$\frac{V_{\text{ave}}}{V_m}$	$\frac{V_{\text{rms}}}{V_m}$	FF	CF
sinusoid	0	$\frac{1}{\sqrt{2}}$	–	$\sqrt{2}$
full-wave rectified sinusoid	$\frac{2}{\pi}$	$\frac{1}{\sqrt{2}}$	$\frac{\pi}{2\sqrt{2}}$	$\sqrt{2}$
half-wave rectified sinusoid	$\frac{1}{\pi}$	$\frac{1}{2}$	$\frac{\pi}{2}$	2
symmetrical square wave	0	1	–	1
unsymmetrical square wave	$\frac{t}{T}$	$\sqrt{\frac{t}{T}}$	$\sqrt{\frac{T}{t}}$	$\sqrt{\frac{T}{t}}$
sawtooth and symmetrical triangular	0	$\frac{1}{\sqrt{3}}$	–	$\sqrt{3}$
sawtooth and unsymmetrical triangular	$\frac{1}{2}$	$\frac{1}{\sqrt{3}}$	$\frac{2}{\sqrt{3}}$	$\sqrt{3}$

18. POWER

- *average power*

$$P_R = \tfrac{1}{2} I_m V_m = \frac{V_m^2}{2R} = I_{\text{rms}} V_{\text{rms}} = IV \qquad 27.44$$

19. REAL POWER AND THE POWER FACTOR

$$P = IV\cos\phi_{\text{pf}} = IV\text{pf} \qquad 27.53$$

$$P = \text{Re}(\mathbf{VI}^*) \qquad 27.55$$

Table 27.3 *Properties of Complex Numbers*

	rectangular form	polar/exponential form
	$\mathbf{Z} = x + jy$	$\mathbf{Z} = \lvert\mathbf{Z}\rvert\angle\theta$ $\mathbf{Z} = \lvert\mathbf{Z}\rvert e^{j\theta} = \lvert\mathbf{Z}\rvert\cos\theta + j\lvert\mathbf{Z}\rvert\sin\theta$
relationship between forms	$x = \lvert\mathbf{Z}\rvert\cos\theta$ $y = \lvert\mathbf{Z}\rvert\sin\theta$	$\lvert\mathbf{Z}\rvert = \sqrt{x^2 + y^2}$ $\theta = \arctan\frac{y}{x}$
complex conjugate	$\mathbf{Z}^* = x - jy$ $\mathbf{ZZ}^* = (x^2 + y^2) = \lvert z\rvert^2$	$\mathbf{Z}^* = \lvert\mathbf{Z}\rvert e^{-j\theta} = \lvert\mathbf{Z}\rvert\angle{-\theta}$ $\mathbf{ZZ}^* = (\lvert\mathbf{Z}\rvert e^{j\theta})(\lvert\mathbf{Z}\rvert e^{-j\theta}) = \lvert\mathbf{Z}\rvert^2$
addition	$\mathbf{Z}_1 + \mathbf{Z}_2 = (x_1 + x_2) + j(y_1 + y_2)$	$\mathbf{Z}_1 + \mathbf{Z}_2 = (\lvert\mathbf{Z}_1\rvert\cos\theta_1 + \lvert\mathbf{Z}_2\rvert\cos\theta_2)$ $+ j(\lvert\mathbf{Z}_1\rvert\sin\theta_1 + \lvert\mathbf{Z}_2\rvert\sin\theta_2)$
multiplication	$\mathbf{Z}_1\mathbf{Z}_2 = (x_1x_2 - y_1y_2) + j(x_1y_2 + x_2y_1)$	$\mathbf{Z}_1\mathbf{Z}_2 = \lvert\mathbf{Z}_1\rvert\lvert\mathbf{Z}_2\rvert\angle\theta_1 + \theta_2$
division	$\frac{\mathbf{Z}_1}{\mathbf{Z}_2} = \frac{(x_1x_2 + y_1y_2) + j(x_2y_1 - x_1y_2)}{\lvert\mathbf{Z}_2\rvert^2}$	$\frac{z_1}{z_2} = \frac{\lvert\mathbf{Z}_1\rvert}{\lvert\mathbf{Z}_2\rvert}\angle\theta_1 - \theta_2$

20. REACTIVE POWER

$$Q = IV\sin\phi_{\text{pf}} \qquad 27.56$$

21. APPARENT POWER

$$S = IV \qquad 27.57$$

22. COMPLEX POWER AND THE POWER TRIANGLE

$$S^2 = P^2 + Q^2 \qquad 27.58$$

Figure 27.11 *Power Triangle*

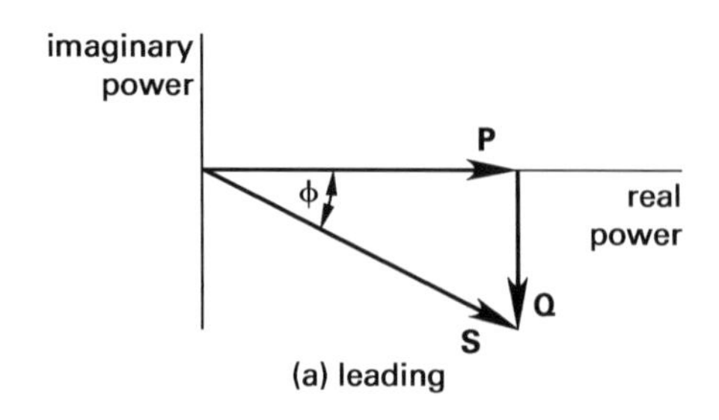

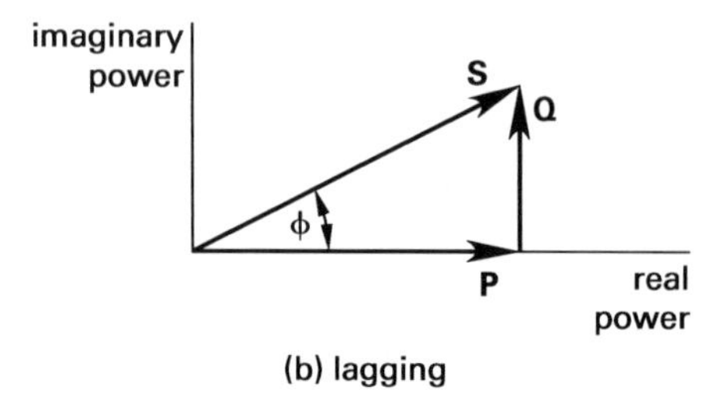

23. MAXIMUM POWER TRANSFER

The conditions for maximum power transfer, and therefore resonance, are given by

$$R_{\text{load}} = R_s \qquad 27.63$$

$$X_{\text{load}} = -X_s \qquad 27.64$$

EPRM Chapter 28 Transformers

2. MAGNETIC COUPLING

$$V_s(t) = \frac{-N_s(2\pi f)\Phi_m\cos\omega t}{\sqrt{2}} = -4.44N_sf\Phi_m\cos\omega t \qquad 28.3$$

Figure 28.1 *Fluxes in a Magnetically Coupled Circuit*

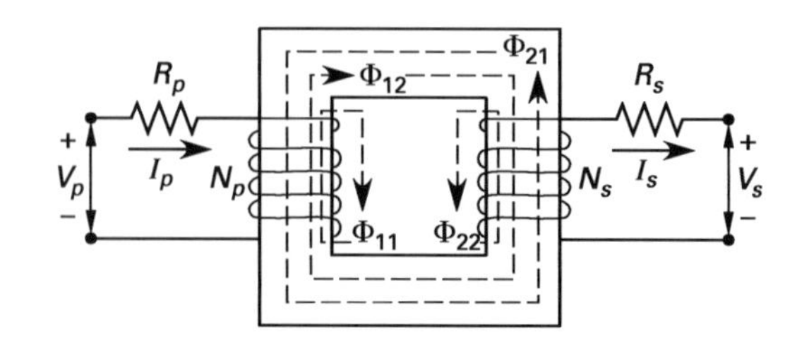

$$
\begin{aligned}
V_p &= I_p(R_p + j\omega L_p) - I_s j\omega M \\
&= I_p(R_p + jX_p) - I_s jX_m \\
&= I_p Z_p - I_s jX_m
\end{aligned}
\qquad 28.7
$$

$$
\begin{aligned}
V_s &= I_s(R_s + j\omega L_s) - I_p j\omega M \\
&= I_s(R_s + jX_s) - I_p jX_m \\
&= I_s Z_s - I_p jX_m
\end{aligned}
\qquad 28.8
$$

Figure 28.2 *Transformer Models*

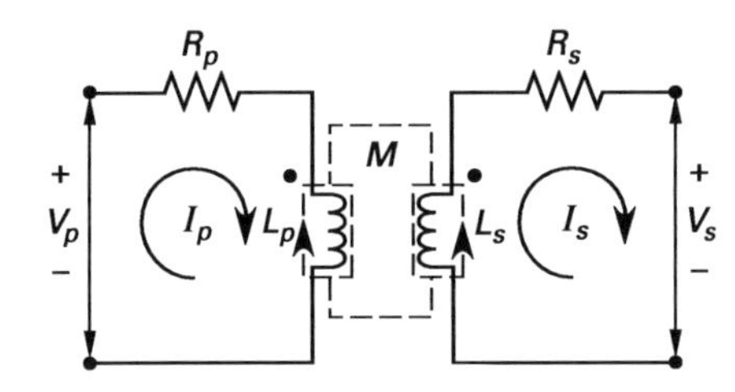

(a) flux linked circuit with mutual inductance

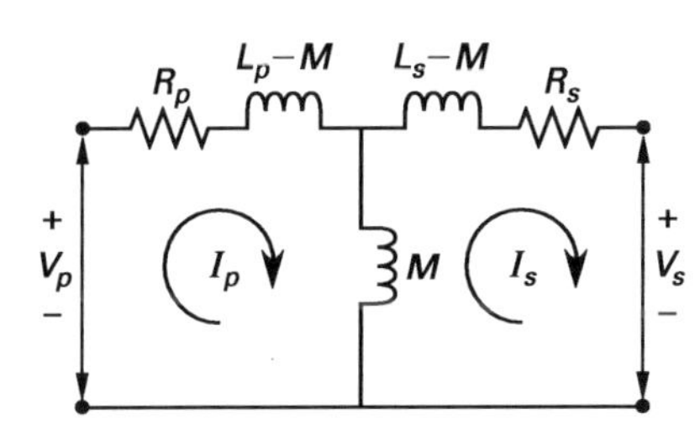

(b) equivalent conductively coupled circuit

Figure 28.4 *Positive and Negative Mutual Inductances*

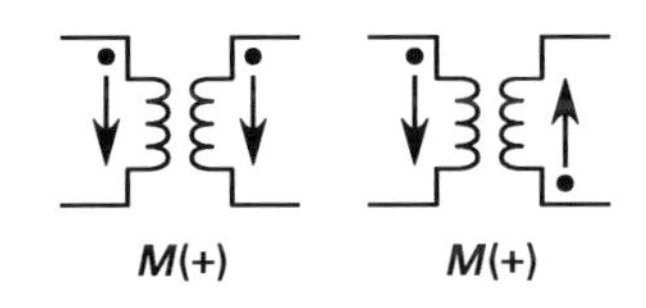

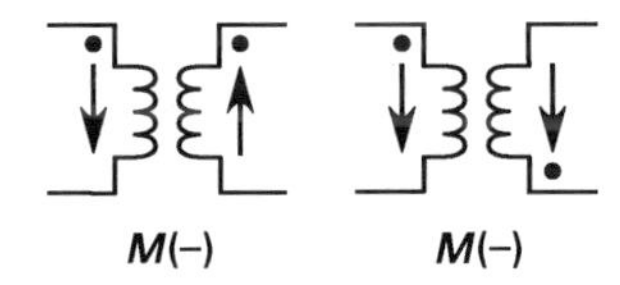

3. IDEAL TRANSFORMERS

$$a = \frac{N_p}{N_s} \qquad 28.10$$

$$I_p V_p = I_s V_s \qquad 28.11$$

$$a = \frac{N_p}{N_s} = \frac{V_p}{V_s} = \frac{I_s}{I_p} = \sqrt{\frac{Z_p}{Z_s}} \qquad 28.12$$

4. IMPEDANCE MATCHING

- *effective primary impedance*, Z_{ep} (or *reflected impedance*, Z_{ref})

$$Z_{\text{ep}} = Z_{\text{ref}} = \frac{V_p}{I_p} = Z_p + a^2 Z_s \qquad 28.14$$

$$a = \sqrt{\frac{Z_p}{Z_s}} \qquad 28.15$$

$$R_p = a^2 R_s \qquad 28.16$$

$$X_p = -a^2 X_s \qquad 28.17$$

5. REAL TRANSFORMERS

Figure 28.8 *Real Transformer Equivalent Circuit*

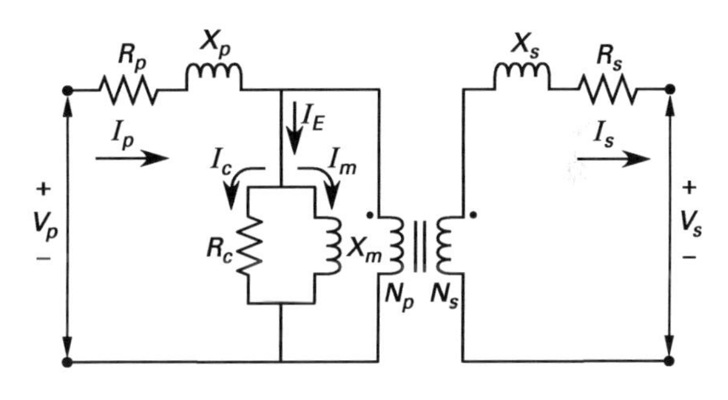

$$X_p = \omega L_p = \frac{V_{X_p}}{I_p} = \frac{4.44 f \Phi_p N_p}{I_p} \qquad 28.18$$

$$X_s = \omega L_s = \frac{V_{X_s}}{I_s} = \frac{4.44 f \Phi_s N_s}{I_s} \qquad 28.19$$

$$a = \sqrt{\frac{L_p}{L_s}} \qquad 28.20$$

6. MAGNETIC HYSTERESIS: BH CURVES

(See Fig. 28.9.)

EPRM Chapter 29
Linear Circuit Analysis

8. LINEAR CIRCUIT ELEMENTS

(See Table 29.1 and Table 29.2.)

9. RESISTANCE

For copper, α_{20} is approximately 3.9×10^{-3} °C^{-1}. The equation for resistivity at any temperature, T (in °C), becomes

$$\rho = \rho_{20}\big(1 + \alpha_{20}(T - 20°\text{C})\big) \qquad 29.3$$

Figure 28.9 Magnetic Hysteresis Loop

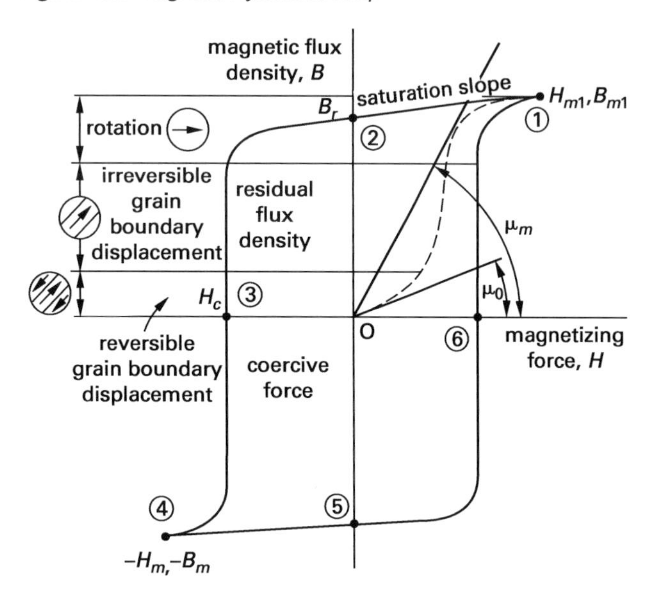

The resistance of a conductor of length l and cross-sectional area A is

$$R = \frac{\rho l}{A} \tag{29.4}$$

- *Ohm's law for AC circuits*

$$V = IR \quad \text{or} \quad \mathbf{V} = \mathbf{I}\mathbf{Z}_R \tag{29.5}$$

- *power dissipated*

$$P = IV = I^2R = \frac{V^2}{R} \tag{29.6}$$

- *equivalent resistance of resistors in series*

$$R_e = R_1 + R_2 + \ldots + R_n \tag{29.7}$$

- *equivalent resistance of resistors in parallel*

$$\frac{1}{R_e} = \frac{1}{R_1} + \frac{1}{R_2} + \ldots + \frac{1}{R_n} \tag{29.8}$$

Table 29.1 Linear Circuit Element Parameters

circuit element	voltage	current	instantaneous power	average power	average energy stored
resistor (+ v −, i →)	$v = iR$	$i = \frac{v}{R}$	$p = iv = i^2R = \frac{v^2}{R}$	$P = IV = I^2R = \frac{V^2}{R}$	–
capacitor (+ v −, i →)	$v = \frac{1}{C}\int i\,dt + \kappa$	$i = C\frac{dv}{dt}$	$p = iv = Cv\frac{dv}{dt}$	0	$U = \frac{1}{2}CV^2$
inductor (+ v −, i →)	$v = L\frac{di}{dt}$	$i = \frac{1}{L}\int v\,dt + \kappa$	$p = iv = Li\frac{di}{dt}$	0	$U = \frac{1}{2}LI^2$

Table 29.2 Linear Circuit Parameters, Time and Frequency Domain Representation

parameter	defining equation	time domain	frequency domain[a]
resistance	$R = \frac{v}{i}$	$v = iR$	$V = IR$
capacitance	$C = \frac{Q}{V}$	$i = C\frac{dv}{dt}$	$I = j\omega CV$
self-inductance	$L = \frac{\Psi}{I}$	$v = L\frac{di}{dt}$	$V = j\omega LI$
mutual inductance[b]	$M_{12} = M_{21} = M = \frac{\Psi_{12}}{I_2} = \frac{\Psi_{21}}{I_1}$	$v_1 = L_1\frac{di_1}{dt} + M\frac{di_2}{dt}$ $v_2 = L_2\frac{di_2}{dt} + M\frac{di_1}{dt}$	$V_1 = j\omega(L_1I_1 + MI_2)$ $V_2 = j\omega(L_2I_2 + MI_1)$

[a]The voltages and currents in the frequency domain column are not shown as vectors; for example, **I** or **V**. They are, however, shown in phasor form. Across any single circuit element (or parameter), the phase angle is determined by the impedance and embodied by the j. If the $j\omega$ (C or L or M) were not shown and instead given as **Z**, the current and voltage would be shown as **I** and **V**, respectively.
[b]Currents I_1 and I_2 (in the mutual inductance row) are in phase. The angle between either I_1 or I_2 and V_1 or V_2 is determined by the impedance angle embodied by the j. Even in a real transformer, this relationship holds, since the equivalent circuit accounts for any phase difference with a magnetizing current, I_m.

10. CAPACITANCE

Capacitance is a measure of the ability to store charge.

$$i = C\frac{dv}{dt} \qquad 29.9$$

For free space, that is, a vacuum, the permittivity is 8.854×10^{-12} F/m.

$$\epsilon = \epsilon_r \epsilon_0 \qquad 29.10$$

- *simple parallel plate capacitor*

$$C = \frac{\epsilon A}{r} \qquad 29.11$$

- *defining equation for capacitance*

$$C = \frac{Q}{V} \qquad 29.12$$

- *(average) energy stored*

$$U = \tfrac{1}{2}CV^2 = \tfrac{1}{2}QV = \tfrac{1}{2}\left(\frac{Q^2}{C}\right) \qquad 29.13$$

- *equivalent capacitance of capacitors in series*

$$\frac{1}{C_e} = \frac{1}{C_1} + \frac{1}{C_2} + \ldots + \frac{1}{C_n} \qquad 29.14$$

- *equivalent capacitance of capacitors in parallel*

$$C_e = C_1 + C_2 + \ldots + C_n \qquad 29.15$$

11. INDUCTANCE

Inductance is a measure of the ability to store magnetic energy.

$$v = L\frac{di}{dt} \qquad 29.16$$

For free space, that is, a vacuum, the permeability is 1.2566×10^{-6} H/m.

$$\mu = \mu_r \mu_0 \qquad 29.17$$

- inductance of a simple toroid (N is the coil turns)

$$L = \frac{\mu N^2 A}{l} \qquad 29.18$$

- *(average) energy stored*

$$U = \tfrac{1}{2}LI^2 = \tfrac{1}{2}\Psi I = \tfrac{1}{2}\left(\frac{\Psi^2}{L}\right) \qquad 29.20$$

- *equivalent inductance of inductors in series*

$$L_e = L_1 + L_2 + \ldots + L_n \qquad 29.21$$

- *equivalent inductance of inductors in parallel*

$$\frac{1}{L_e} = \frac{1}{L_1} + \frac{1}{L_2} + \ldots + \frac{1}{L_n} \qquad 29.22$$

12. MUTUAL INDUCTANCE

(See Table 29.2.)

$$U = \tfrac{1}{2}L_1 I_1^2 + \tfrac{1}{2}L_2 I_2^2 + M I_1 I_2 \qquad 29.23$$

17. DELTA-WYE TRANSFORMATIONS

$$Z_a = \frac{Z_1Z_2 + Z_1Z_3 + Z_2Z_3}{Z_3} \qquad 29.26$$

$$Z_b = \frac{Z_1Z_2 + Z_1Z_3 + Z_2Z_3}{Z_1} \qquad 29.27$$

$$Z_c = \frac{Z_1Z_2 + Z_1Z_3 + Z_2Z_3}{Z_2} \qquad 29.28$$

$$Z_1 = \frac{Z_aZ_c}{Z_a + Z_b + Z_c} \qquad 29.29$$

$$Z_2 = \frac{Z_aZ_b}{Z_a + Z_b + Z_c} \qquad 29.30$$

$$Z_3 = \frac{Z_bZ_c}{Z_a + Z_b + Z_c} \qquad 29.31$$

Figure 29.10 *Delta (Pi)-Wye (T) Configurations*

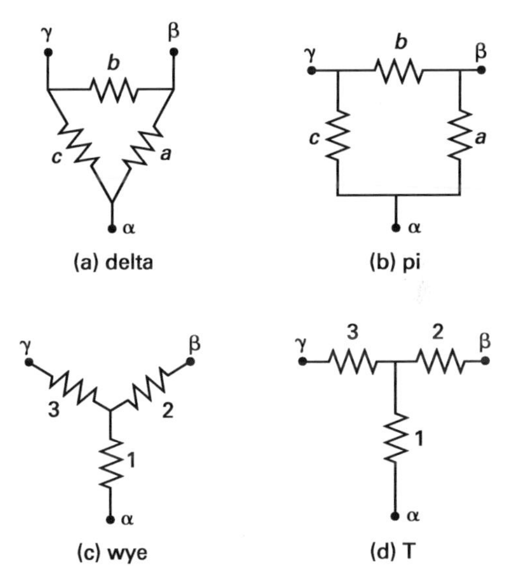

18. THEVENIN'S THEOREM

$$\mathbf{V}_{\text{Th}} = \mathbf{V}_{\text{oc}} \qquad 29.32$$

$$\mathbf{Z}_{\text{Th}} = \frac{\mathbf{V}_{\text{oc}}}{\mathbf{I}_{\text{sc}}} \qquad 29.33$$

(See Fig. 29.11.)

19. NORTON'S THEOREM

$$\mathbf{I}_N = \mathbf{I}_{\text{sc}} \qquad 29.34$$

$$\mathbf{Z}_N = \frac{\mathbf{V}_{\text{oc}}}{\mathbf{I}_{\text{sc}}} \qquad 29.35$$

(See Fig. 29.12.)

Figure 29.11 Thevenin Equivalent Circuit

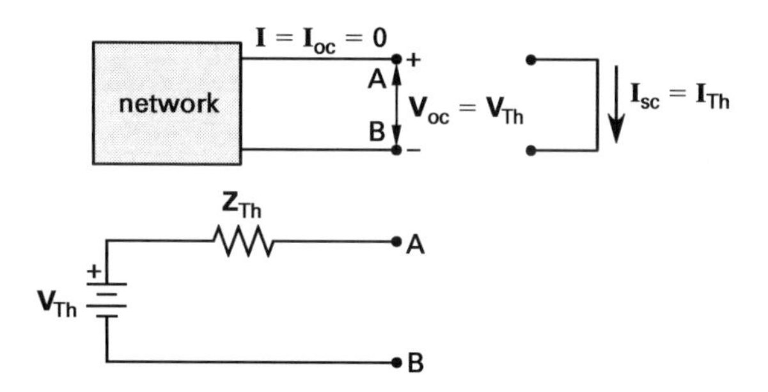

Figure 29.12 Norton Equivalent Circuit

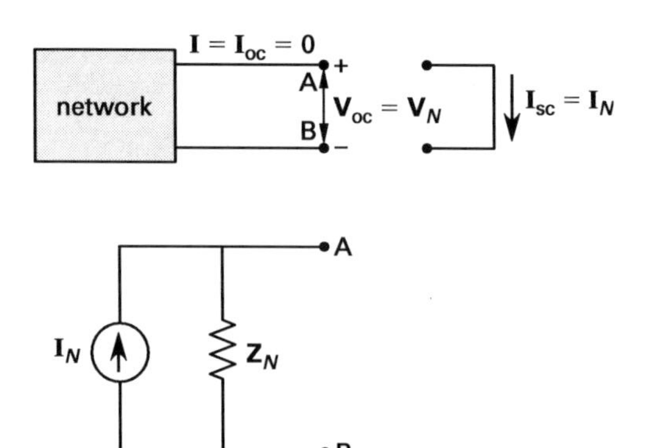

20. MAXIMUM POWER TRANSFER THEOREM

Where the load impedance varies and the source impedance is fixed, maximum power transfer occurs when the load and source impedances are complex conjugates. That is, $\mathbf{Z}_l = \mathbf{Z}_s^*$ or $R_{\text{load}} + jX_{\text{load}} = R_s - jX_s$.

21. SUPERPOSITION THEOREM

step 1: Replace all sources except one by their internal resistances. Ideal current sources are replaced by open circuits. Ideal voltage sources are replaced by short circuits.

step 2: Compute the desired quantity, either voltage or current caused by the single source, for the element in question.

step 3: Repeat steps 1 and 2 for each of the sources in turn.

step 4: Sum the calculated values obtained for the current or voltage obtained in step 2. The result is the actual value of the current or voltage in the element for the complete circuit.

Superposition is not valid for circuits in which the following conditions exist.

- The capacitors have an initial charge (i.e., an initial voltage) not equal to zero.
- The inductors have an initial magnetic field (i.e., an initial current) not equal to zero.
- Dependent sources are used.

24. KIRCHHOFF'S VOLTAGE LAW

step 1: Identify the loop.

step 2: Pick a loop direction.

step 3: Assign the loop current in the direction picked in step 2.

step 4: Assign voltage polarities consistent with the loop current direction in step 3.

step 5: Apply KVL to the loop using Ohm's law to express the voltages across each circuit element.

step 6: Solve the equation for the desired quantity.

25. KIRCHHOFF'S CURRENT LAW

step 1: Identify the nodes and pick a reference or datum node.

step 2: Label the node-to-datum voltage for each unknown node.

step 3: Pick a current direction for each path at every node.

step 4: Apply KCL to the nodes using Ohm's law to express the currents through each circuit branch.

step 5: Solve the equations for the desired quantity.

26. LOOP ANALYSIS

step 1: Select $n-1$ loops, that is, one loop less than the total number of possible loops.

step 2: Assign current directions for the selected loops. Show the direction of the current with an arrow.

step 3: Write Kirchhoff's voltage law for each of the selected loops. Assign polarities based on the direction of the loop current. Where two loop currents flow through an element, they are summed to determine the voltage drop in that element.

step 4: Solve the $n-1$ equations from step 3 for the unknown currents.

step 5: If required, determine the actual current in an element by summing the loop currents flowing through the element.

27. NODE ANALYSIS

step 1: Simplify the circuit, if possible, by combining resistors in series or parallel or by combining current sources in parallel. Identify all nodes. The minimum number of equations required will be $n-1$ where n represents the number of principal nodes.

step 2: Choose one node as the reference node, that is, the node that will be assumed to have ground potential (0 V). To minimize the number of terms in the equations, select the node with the largest number of circuit elements to serve as the reference node.

step 3: Write Kirchhoff's current law for each principal node except the reference node, which is assumed to have a zero potential.

step 4: Solve the $n-1$ equations from step 3 to determine the unknown voltages.

step 5: If required, use the calculated node voltages to determine any branch current desired.

29. VOLTAGE AND CURRENT DIVIDERS

$$\mathbf{V}_2 = \mathbf{V}_s \left(\frac{\mathbf{Z}_2}{\mathbf{Z}_1 + \mathbf{Z}_2}\right) \qquad 29.38$$

$$\mathbf{I}_2 = \mathbf{I}_s \left(\frac{\mathbf{Z}_1}{\mathbf{Z}_1 + \mathbf{Z}_2}\right) = \mathbf{I}_s \left(\frac{\mathbf{G}_2}{\mathbf{G}_1 + \mathbf{G}_2}\right) \qquad 29.40$$

Figure 29.14 *Divider Circuits*

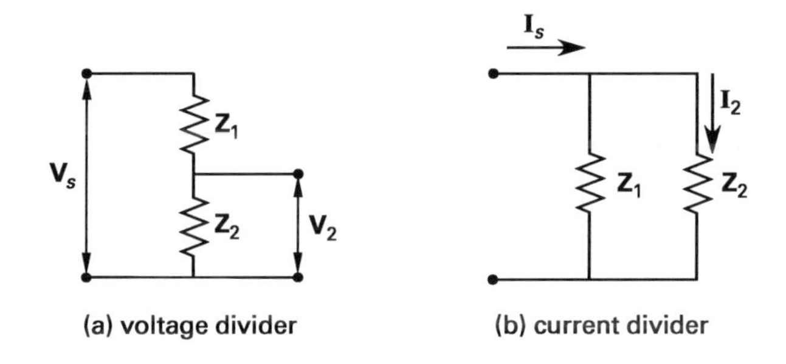

30. STEADY-STATE AND TRANSIENT IMPEDANCE ANALYSIS

DC Steady-State Impedance Analysis

- *resistance*

$$Z_R|_{\mathrm{DC}} = R \qquad 29.41$$

- *inductance*

$$v = L\,\frac{di}{dt} = L(0) = 0 \qquad 29.42$$

$$Z = \frac{v}{i} = \frac{0}{i} = 0 \qquad 29.43$$

$$Z_L|_{\mathrm{DC}} = 0 \quad \text{[short circuit]} \qquad 29.44$$

- *capacitance*

$$i = C\,\frac{dv}{dt} = C(0) = 0 \qquad 29.45$$

$$Z = \frac{v}{i} = \frac{v}{0} \rightarrow \infty \qquad 29.46$$

$$Z_C|_{\mathrm{DC}} = \infty \quad \text{[open circuit]} \qquad 29.47$$

AC Steady-State Impedance Analysis

- *resistance*

$$Z_R|_{\mathrm{AC}} = R \qquad 29.48$$

- *inductance*

$$v = L\,\frac{di}{dt} = Lj\omega i \qquad 29.49$$

$$Z = \frac{v}{i} = \frac{Lj\omega i}{i} = j\omega L \qquad 29.50$$

$$Z_L|_{\mathrm{AC}} = j\omega L \qquad 29.51$$

- *capacitance*

$$i = C\,\frac{dv}{dt} = Cj\omega v \qquad 29.52$$

$$Z = \frac{v}{i} = \frac{v}{Cj\omega v} = \frac{1}{j\omega C} \qquad 29.53$$

$$Z_C|_{\mathrm{AC}} = \frac{1}{j\omega C} \qquad 29.54$$

Transient Impedance Analysis

Transient impedance analysis is based on the phasor form with the complex variable s substituted for $j\omega$. The variable $s = \sigma + j\omega$ and is the same as the Laplace transform variable.

- *resistance*

$$Z_R = R \qquad 29.55$$

- *inductance*

$$Z_L = sL \qquad 29.56$$

- *capacitance*

$$Z_C = \frac{1}{sC} \qquad 29.57$$

31. TWO-PORT NETWORKS

Figure 29.15 *Two-Port Network*

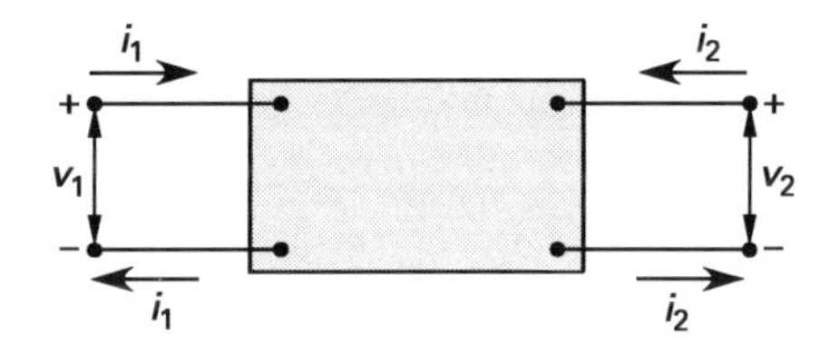

(See Table 29.3.)

EPRM Chapter 30
Transient Analysis

1. FUNDAMENTALS

(See Table 30.1.)

$$f(t) = b\left(\frac{dx}{dt}\right) + cx \qquad 30.1$$

The solution to Eq. 30.1 is of the general form

$$x(t) = \kappa + Ae^{-t/\tau} \qquad 30.2$$

Table 29.3 *Two-Port Network Parameters*

representation		deriving equations		
impedance $\begin{bmatrix} z_{11} & z_{12} \\ z_{21} & z_{22} \end{bmatrix}$	$z_{11} = \frac{V_1}{I_1}$ $I_2 = 0$	$z_{12} = \frac{V_1}{I_2}$ $I_1 = 0$	$z_{21} = \frac{V_2}{I_1}$ $I_2 = 0$	$z_{22} = \frac{V_2}{I_2}$ $I_1 = 0$
admittance $\begin{bmatrix} y_{11} & y_{12} \\ y_{21} & y_{22} \end{bmatrix}$	$y_{11} = \frac{I_1}{V_1}$ $V_2 = 0$	$y_{12} = \frac{I_1}{V_2}$ $V_1 = 0$	$y_{21} = \frac{I_2}{V_1}$ $V_2 = 0$	$y_{22} = \frac{I_2}{V_2}$ $V_1 = 0$
hybrid $\begin{bmatrix} h_{11} & h_{12} \\ h_{21} & h_{22} \end{bmatrix}$	$h_{11} = \frac{V_1}{I_1}$ $V_2 = 0$	$h_{12} = \frac{V_1}{V_2}$ $I_1 = 0$	$h_{21} = \frac{I_2}{I_1}$ $V_2 = 0$	$h_{22} = \frac{I_2}{V_2}$ $I_1 = 0$
inverse hybrid $\begin{bmatrix} g_{11} & g_{12} \\ g_{21} & g_{22} \end{bmatrix}$	$g_{11} = \frac{I_1}{V_1}$ $I_2 = 0$	$g_{12} = \frac{I_1}{I_2}$ $V_1 = 0$	$g_{21} = \frac{V_2}{V_1}$ $I_2 = 0$	$g_{22} = \frac{V_2}{I_2}$ $V_1 = 0$
transmission or chain $\begin{bmatrix} A & B \\ C & D \end{bmatrix}$	$A = \frac{V_1}{V_2}$ $I_2 = 0$	$B = \frac{-V_1}{I_2}$ $V_2 = 0$	$C = \frac{I_1}{V_2}$ $I_2 = 0$	$D = \frac{-I_1}{I_2}$ $V_2 = 0$
inverse transmission $\begin{bmatrix} \alpha & \beta \\ \gamma & \delta \end{bmatrix}$	$\alpha = \frac{V_2}{V_1}$ $I_1 = 0$	$\beta = \frac{-V_2}{I_1}$ $V_1 = 0$	$\gamma = \frac{I_2}{V_1}$ $I_1 = 0$	$\delta = \frac{-I_2}{I_1}$ $V_1 = 0$

- *time constant*

$$\tau = \frac{b}{c} \qquad 30.3$$

- *time constant for an RC circuit*

$$\tau = RC \qquad 30.4$$

- *time constant for an RL circuit*

$$\tau = \frac{L}{R} \qquad 30.5$$

10. RESONANT CIRCUITS

A *resonant circuit* has a zero current phase angle difference. The frequency at which the circuit becomes purely resistive is the *resonant frequency.*

The *quality factor*, Q, for a circuit is a dimensionless ratio that compares the reactive energy stored in an inductor each cycle to the resistive energy dissipated.

$$\begin{aligned} Q &= 2\pi\left(\frac{\text{maximum energy stored per cycle}}{\text{energy dissipated per cycle}}\right) \\ &= \frac{f_0}{(\text{BW})_{\text{Hz}}} = \frac{\omega_0}{(\text{BW})_{\text{rad/s}}} \\ &= \frac{f_0}{f_2 - f_1} = \frac{\omega_0}{\omega_2 - \omega_1} \quad \begin{bmatrix}\text{parallel} \\ \text{or series}\end{bmatrix} \end{aligned} \qquad 30.25$$

- *energy stored in the inductor of a series RLC circuit each cycle*

$$U = \frac{I_m^2 L}{2} = I^2 L = Q\left(\frac{I^2 R}{2\pi f_0}\right) \qquad 30.26$$

Figure 30.9 *Series Resonance (Band-Pass Filter)*

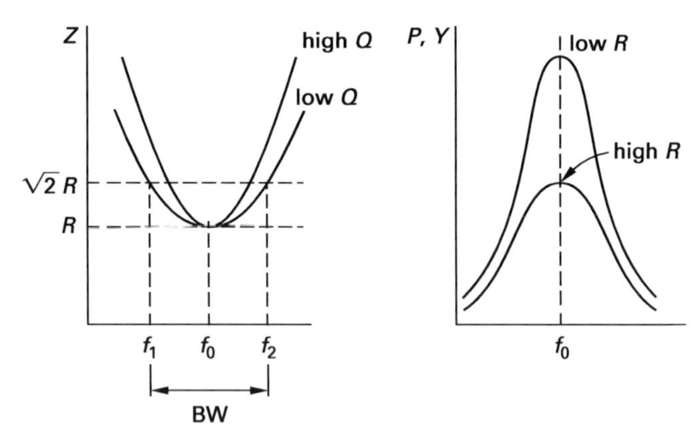

Figure 30.10 *Parallel Resonance (Band-Reject Filter)*

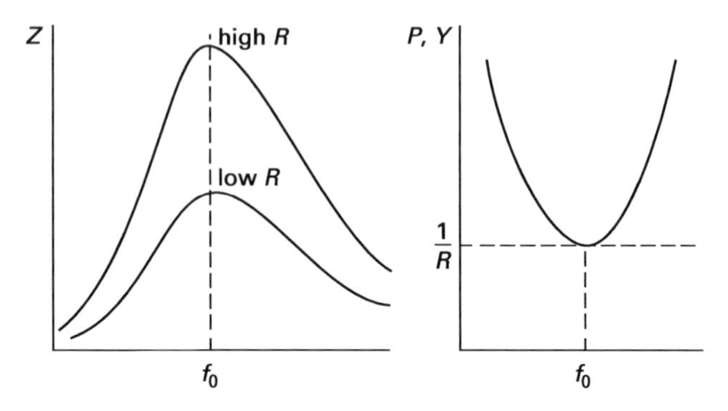

***Table 30.1** Transient Response*

type of circuit	response
series RC, charging	
$\tau = RC$ $e^{-N} = e^{-t/\tau} = e^{-t/RC}$	$V_{\text{bat}} = v_R(t) + v_C(t)$ $i(t) = \left(\frac{V_{\text{bat}} - V_0}{R}\right)e^{-N}$ $v_R(t) = i(t)R = (V_{\text{bat}} - V_0)e^{-N}$ $v_C(t) = V_0 + (V_{\text{bat}} - V_0) \times (1 - e^{-N})$ $Q_C(t) = C\left(V_0 + (V_{\text{bat}} - V_0) \times (1 - e^{-N})\right)$
series RC, discharging	
$\tau = RC$ $e^{-N} = e^{-t/\tau} = e^{-t/RC}$	$0 = v_R(t) + v_C(t)$ $i(t) = \left(\frac{V_0}{R}\right)e^{-N}$ $v_R(t) = -V_0e^{-N}$ $v_C(t) = V_0e^{-N}$ $Q_C(t) = CV_0e^{-N}$
series RL, charging	
$\tau = \frac{L}{R}$ $e^{-N} = e^{-t/\tau} = e^{-t/(L/R)}$	$V_{\text{bat}} = v_R(t) + v_L(t)$ $i(t) = I_0e^{-N} + \left(\frac{V_{\text{bat}}}{R}\right)(1 - e^{-N})$ $v_R(t) = i(t)R = I_0Re^{-N} + V_{\text{bat}}(1 - e^{-N})$ $v_L(t) = (V_{\text{bat}} - I_0R)e^{-N}$
series RL, discharging	
$\tau = \frac{L}{R}$ $e^{-N} = e^{-t/\tau} = e^{-t/(L/R)}$	$0 = v_R(t) + v_L(t)$ $i(t) = I_0e^{-N}$ $v_R(t) = I_0Re^{-N}$ $v_L(t) = -I_0Re^{-N}$

- *relationships between the half-power points and quality factor*

$$f_1,\ f_2 = f_0\left(\sqrt{1 + \frac{1}{4Q^2}} \mp \frac{1}{2Q}\right) \approx f_0 \mp \frac{f_0}{2Q} = f_0 \mp \frac{\text{BW}}{2} \qquad 30.27$$

11. SERIES RESONANCE

- *quality factor for a series RLC circuit*

$$Q = \frac{X}{R} = \frac{\omega_0 L}{R} = \frac{1}{\omega_0 RC} = \frac{1}{R}\sqrt{\frac{L}{C}} = \frac{\omega_0}{(\text{BW})_{\text{rad/s}}} = \frac{f_0}{(\text{BW})_{\text{Hz}}} = G\omega_0 L = \frac{G}{\omega_0 C} \qquad 30.32$$

12. PARALLEL RESONANCE

At resonance,

$$X_L = X_C \qquad 30.33$$

$$\omega_0 L = \frac{1}{\omega_0 C} \qquad 30.34$$

$$\omega_0 = 2\pi f_0 = \frac{1}{\sqrt{LC}} \qquad 30.35$$

- *quality factor for a parallel RLC circuit*

$$Q = \frac{R}{X} = \omega_0 RC = \frac{R}{\omega_0 L} = R\sqrt{\frac{C}{L}} = \frac{\omega_0}{(\text{BW})_{\text{rad/s}}} = \frac{f_0}{(\text{BW})_{\text{Hz}}} = \frac{\omega_0 C}{G} = \frac{1}{G\omega_0 L} \qquad 30.37$$

EPRM Chapter 31
Time Response

2. FIRST-ORDER ANALYSIS

$$f(t) = b\frac{dx}{dt} + cx \qquad 31.2$$

$$x(t) = \kappa + Ae^{-t/\tau} \qquad 31.3$$

A capacitive first-order circuit generally uses voltage as the dependent variable.

$$v_{\text{Th}} = R_{\text{Th}}C\left(\frac{dv_C}{dt}\right) + v_C \qquad 31.5$$

The solution is

$$v_C(t) = V_{C,\text{ss}} + Ae^{-t/\tau} \quad \text{31.6}$$

An inductive first-order circuit generally uses current as the dependent variable.

$$v_{\text{Th}} = L\left(\frac{di_L}{dt}\right) + R_{\text{Th}} i_L \quad \text{31.8}$$

The solution is

$$i_L(t) = I_{L,\text{ss}} + Ae^{-t/\tau} \quad \text{31.9}$$

5. SECOND-ORDER ANALYSIS

The differential equation for second-order circuits takes on the general form

$$f(t) = a\frac{d^2x}{dt^2} + b\frac{dx}{dt} + cx \quad \text{31.13}$$

The associated characteristic equation, written with the roots shown in the s domain, is

$$as^2 + bs + c = 0 \quad \text{31.14}$$

6. SECOND-ORDER ANALYSIS: OVERDAMPED

If the two roots of Eq. 31.14 are real and different from one another, equivalent to the condition $b^2 > 4ac$, the solution is of the form

$$x(t) = \kappa + Ae^{s_1 t} + Be^{s_2 t} \quad \text{31.15}$$

$$s_1 = \frac{-b + \sqrt{b^2 - 4ac}}{2a} \quad \text{31.16}$$

$$s_2 = \frac{-b - \sqrt{b^2 - 4ac}}{2a} \quad \text{31.17}$$

$$A = \tfrac{1}{2}\left(1 + \frac{b}{\sqrt{b^2 - 4ac}}\right)(x(0^+) - x_{\text{ss}}) + \left(\frac{a}{\sqrt{b^2 - 4ac}}\right)(x'(0^+) - x'_{\text{ss}}) \quad \text{31.24}$$

$$B = \tfrac{1}{2}\left(1 - \frac{b}{\sqrt{b^2 - 4ac}}\right)(x(0^+) - x_{\text{ss}}) - \left(\frac{a}{\sqrt{b^2 - 4ac}}\right)(x'(0^+) - x'_{\text{ss}}) \quad \text{31.25}$$

7. SECOND-ORDER ANALYSIS: CRITICALLY DAMPED

If the two roots of Eq. 31.14 are real and the same, equivalent to the condition $b^2 = 4ac$, the solution is of the form

$$x(t) = \kappa + Ae^{st} + Bte^{st} \quad \text{31.26}$$

$$s = \frac{-b}{2a} \quad \text{31.27}$$

$$A = x(0^+) - x_{\text{ss}} \quad \text{31.34}$$

$$B = x'(0^+) - x'_{\text{ss}} + \left(\frac{b}{2a}\right)A \quad \text{31.35}$$

8. SECOND-ORDER ANALYSIS: UNDERDAMPED

If the two roots of Eq. 31.14 are complex conjugates, equivalent to the condition $b^2 < 4ac$, the solution is of the form

$$x(t) = \kappa + e^{\alpha t}\left(A_e e^{j\beta t} + B_e e^{-j\beta t}\right) \quad \text{31.36}$$

This can also be represented in the sinusoidal form as

$$x(t) = \kappa + e^{\alpha t}\left(A\cos\beta t + B\sin\beta t\right) \quad \text{31.37}$$

$$s_1 = \alpha + j\beta \quad \text{31.38}$$

$$s_2 = \alpha - j\beta \quad \text{31.39}$$

The roots of the characteristic equation are

$$\alpha = \frac{-b}{2a} \quad \text{31.40}$$

$$\beta = \frac{\sqrt{4ac - b^2}}{2a} \quad \text{31.41}$$

$$A = x(0^+) - x_{\text{ss}} \quad \text{31.48}$$

$$B = \left(\frac{-\alpha}{\beta}\right)A + \left(\frac{1}{\beta}\right)(x'(0^+) - x'_{\text{ss}}) \quad \text{31.49}$$

11. COMPLEX FREQUENCY

$$\mathbf{s} = \sigma + j\omega$$

Table 31.4 *Generalized Impedance in the s-Domain*

impedance type	frequency domain value	*s*-domain value
Z_R	R	R
Z_L	$j\omega L$	sL
Z_C	$\frac{1}{j\omega C}$	$\frac{1}{sC}$

12. LAPLACE TRANSFORM ANALYSIS

- *Laplace transform*

$$\mathcal{L}\{f(t)\} = \int_{0+}^{\infty} f(t)e^{-st}dt \quad \text{31.76}$$

$$\mathcal{L}\{f(t)\} = F(s) \quad \text{31.77}$$

$$\mathcal{L}\left\{\frac{d^2f(t)}{dt^2}\right\} = s^2F(s) - sf(0^+) - f'(0^+) \quad \text{31.78}$$

$$\mathcal{L}\left\{\frac{df(t)}{dt}\right\} = sF(s) - f(0^+) \quad \text{31.79}$$

- *inverse Laplace transform*

$$\mathcal{L}^{-1}\{F(s)\} = f(t) = \frac{1}{2\pi j}\int_{\sigma-j\infty}^{\sigma+j\infty} F(s)e^{st}dt \quad \text{31.80}$$

The classical method is useful when the desired quantity is one of the state variables, capacitor voltage or inductor current. If the desired variables are other than the state variables, the *circuit transformation method* is the most direct.

13. CAPACITANCE IN THE s-DOMAIN

$$\mathcal{L}\{v_C(t)\} = V_C(s) \quad 31.81$$

$$\mathcal{L}\{i_C(t)\} = \mathcal{L}\left\{C\frac{dv_C(t)}{dt}\right\} = sCV_C(s) - Cv_C(0) \quad 31.82$$

14. INDUCTANCE IN THE s-DOMAIN

Figure 31.12 Inductor Impedance and Initial Current Source Model

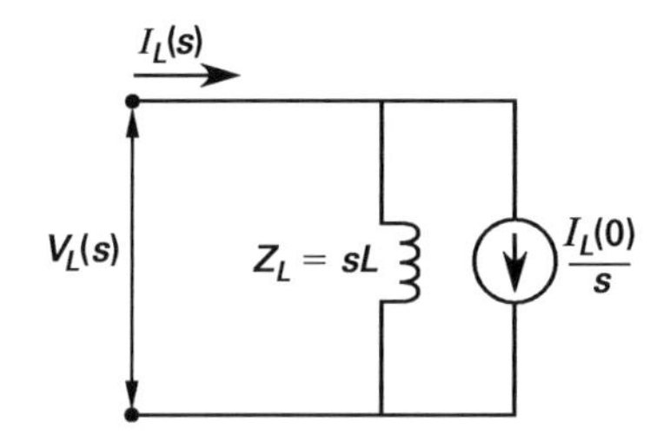

EPRM Chapter 32
Frequency Response

2. TRANSFER FUNCTION

The term *transfer function* refers to the relationship of one electrical parameter in a network to a second electrical parameter elsewhere in the network.

$$T_{\text{net}}(s) = \frac{Y(s)}{X(s)} = A\left(\frac{(s-z_1)(s-z_2)\cdots(s-z_n)}{(s-p_1)(s-p_2)\cdots(s-p_d)}\right) \quad 32.2$$

The complex constants represented as $z_y(y=1, 2, \ldots, n)$ are the *zeros* of $T_{\text{net}}(s)$, and are plotted in the s-domain as circles. The complex constants represented as p_x $(x=1, 2, \ldots, d)$ are the *poles* of $T_{\text{net}}(s)$ and are plotted in the s-domain as X's.

3. STEADY-STATE RESPONSE

$$(s - z_n) = N_n\angle\alpha_n \quad 32.3$$

$$(s - p_d) = D_d\angle\beta_d \quad 32.4$$

$$T_{\text{net}}(s) = A\left(\frac{(N_1\angle\alpha_1)(N_2\angle\alpha_2)\cdots(N_n\angle\alpha_n)}{(D_1\angle\beta_1)(D_2\angle\beta_2)\cdots(D_d\angle\beta_d)}\right) = A\left(\frac{N_1N_2\cdots N_n}{D_1D_2\cdots D_d}\right)\angle\left(\begin{array}{c}\alpha_1+\alpha_2+\cdots\\ +\alpha_n\end{array}\right) - (\beta_1+\beta_2+\cdots+\beta_d) \quad 32.5$$

4. TRANSIENT RESPONSE

The transient response is characterized by the poles of the network function.

6. BODE PLOT PRINCIPLES: MAGNITUDE PLOT

A *Bode plot* or *Bode diagram* is a plot of the gain or phase of an electrical device or network against the frequency.

- *gain* (any transfer function's magnitude)

$$G = 20\log|T(s)| \quad 32.19$$

- *transfer function for a single zero at the origin*

$$T(s) = \sigma_0\left(\frac{s}{\sigma_0}\right) \quad 32.20$$

$$G = 20\log\left|\sigma_0\left(\frac{j\omega}{\sigma_0}\right)\right| = 20\log\sigma_0 + 20\log\frac{\omega}{\sigma_0} \quad 32.21$$

Figure 32.5 Bode Plot: Single Zero at Origin

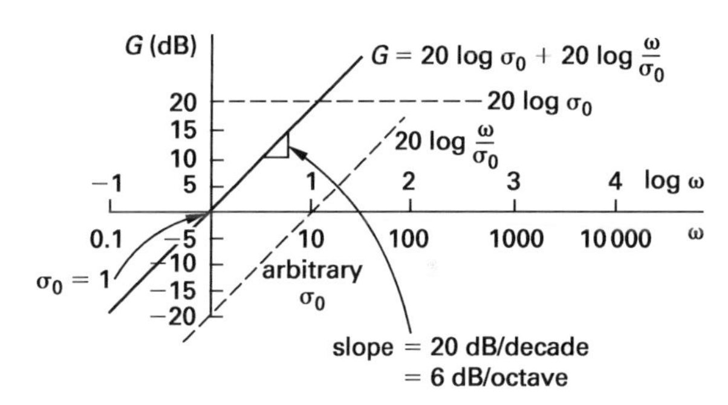

- *transfer function for a single pole at the origin*

$$T(s) = \frac{1}{\sigma_0\left(\frac{s}{\sigma_0}\right)} \quad 32.24$$

- *gain with $s = j\omega$*

$$G = 20\log\left|\frac{1}{\sigma_0\left(\frac{j\omega}{\sigma_0}\right)}\right| = -20\log\sigma_0 - 20\log\frac{\omega}{\sigma_0} \quad 32.25$$

- *transfer function for a single zero on the negative real axis at $-\sigma_0$ in the s plane*

$$T(s) = \sigma_0\left(1 + \frac{s}{\sigma_0}\right) \quad 32.26$$

Figure 32.6 Bode Plot: Single Pole at Origin

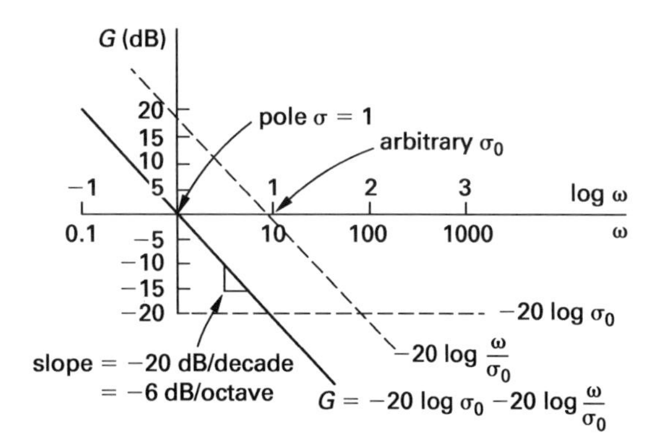

- *gain with* $s = j\omega$

$$G = 20\log\left|\sigma_0\left(1+\frac{j\omega}{\sigma_0}\right)\right|$$

$$= 20\log\sigma_0 + 20\log\left|\left(1+\frac{j\omega}{\sigma_0}\right)\right| \qquad 32.27$$

Figure 32.7 Bode Plot: Single Zero on Negative Real Axis

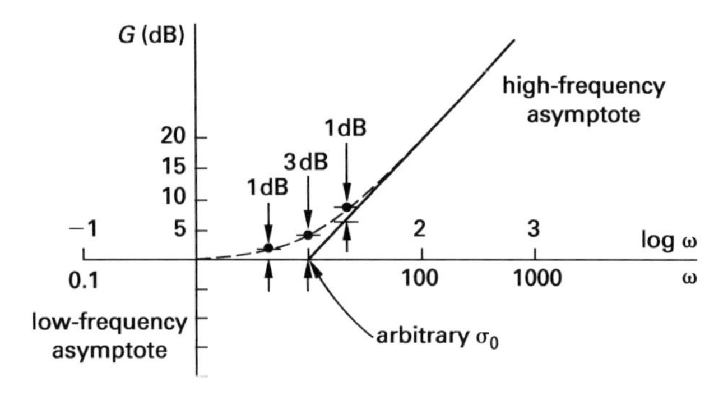

- *transfer function for a single pole on the negative real axis at* $-\sigma_0$ *in the s plane*

$$T(s) = \frac{1}{s+\sigma_d}$$

- *gain with* $s = jw$

$$G = 20\log\left|\left(\frac{1}{\sigma_0\left(1+\frac{j\omega}{\sigma_0}\right)}\right)\right|$$

$$= -20\log\sigma_0 - 20\log\left|\left(1+\frac{j\omega}{\sigma_0}\right)\right| \qquad 32.32$$

Figure 32.8 Bode Plot: Single Pole on Negative Real Axis

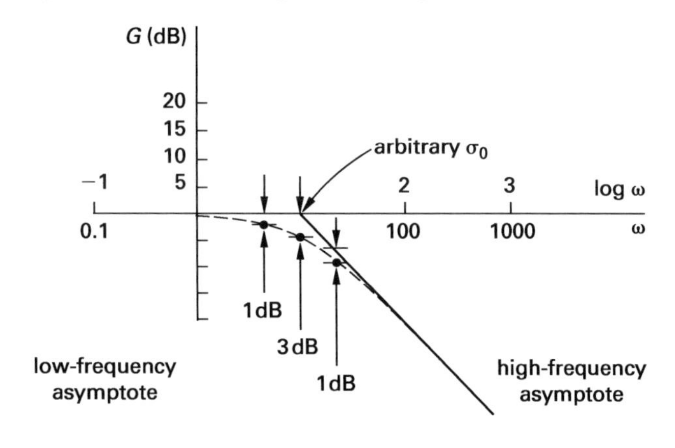

7. BODE PLOT PRINCIPLES: PHASE PLOT

$$\theta_z = \arctan\frac{\omega}{\sigma_{zn}} \quad \text{[radians]} \qquad 32.35$$

$$\theta_p = -\arctan\frac{\omega}{\sigma_{pd}} \quad \text{[radians]} \qquad 32.36$$

Figure 32.9 Bode Phase Plot for a Zero or Pole

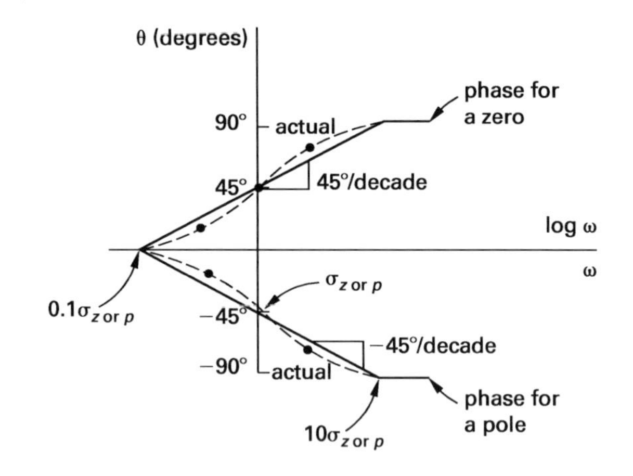

8. BODE PLOT METHODS

- *general form of the transfer function of a circuit*

$$T(s) = A\left(\frac{(s+z_1)(s+z_2)\cdots\times(s+z_n)(s)^y}{(s+p_1)(s+p_2)\cdots\times(s+p_d)(s)^x}\right) \qquad 32.39$$

Using the normalized notation changes Eq. 32.39 to Eq. 32.40.

$$T(j\omega) = A\left(\frac{z_1z_2\cdots z_n}{p_1p_2\cdots p_d}\right) \times \left(\frac{\left(1+\frac{j\omega}{z_1}\right)\left(1+\frac{j\omega}{z_2}\right)\cdots\times\left(1+\frac{j\omega}{z_n}\right)(j\omega)^y}{\left(1+\frac{j\omega}{p_1}\right)\left(1+\frac{j\omega}{p_2}\right)\cdots\times\left(1+\frac{j\omega}{p_d}\right)(j\omega)^x}\right) \qquad 32.40$$

Combining the constant term A with the values of the zeros and poles gives a new constant, K.

$$T(j\omega) = K\left(\frac{\left(1+\frac{j\omega}{z_1}\right)\left(1+\frac{j\omega}{z_2}\right)\cdots\times\left(1+\frac{j\omega}{z_n}\right)(j\omega)^y}{\left(1+\frac{j\omega}{p_1}\right)\left(1+\frac{j\omega}{p_2}\right)\cdots\times\left(1+\frac{j\omega}{p_d}\right)(j\omega)^x}\right) \qquad 32.41$$

- *constant K*

$$K = A\frac{\prod_1^n z_n}{\prod_1^d p_d} \qquad 32.42$$

$$G = 20\log K + \sum_{n=1}^{n} 20\log\left|\left(1+\frac{j\omega}{z_n}\right)\right| + 20y\log\omega - \sum_{d=1}^{d} 20\log\left|1+\frac{j\omega}{p_d}\right| - 20x\log\omega \qquad 32.43$$

- *angle of the gain, θ_G, in degrees*

$$\theta_G = \sum_{n=1}^{n}\arctan\frac{\omega}{z_n} + 90y - \sum_{d=1}^{d}\arctan\frac{\omega}{p_d} - 90x \qquad 32.44$$

Generation

EPRM Chapter 33
Generation Systems

6. ALTERNATING CURRENT GENERATORS

$$n_s = \frac{120f}{p} \quad 33.5$$

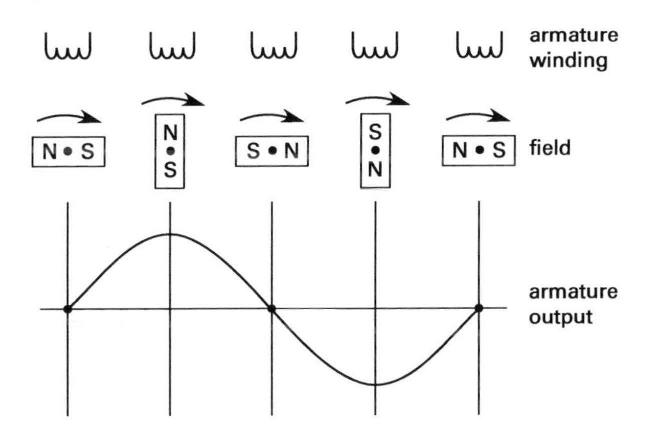

Figure 33.8 AC Output Generation

7. PARALLEL OPERATION

$$P = \frac{f_{\text{nl}} - f_{\text{sys}}}{f_{\text{droop}}} \quad 33.8$$

$$Q = \frac{V_{\text{sys}} - V_{\text{nl}}}{V_{\text{droop}}} \quad 33.9$$

If the droop is given or used as a positive value, Eq. 33.9 becomes

$$Q = \frac{V_{\text{nl}} - V_{\text{sys}}}{V_{\text{droop}}} \quad 33.10$$

Figure 33.9 Real Load Sharing

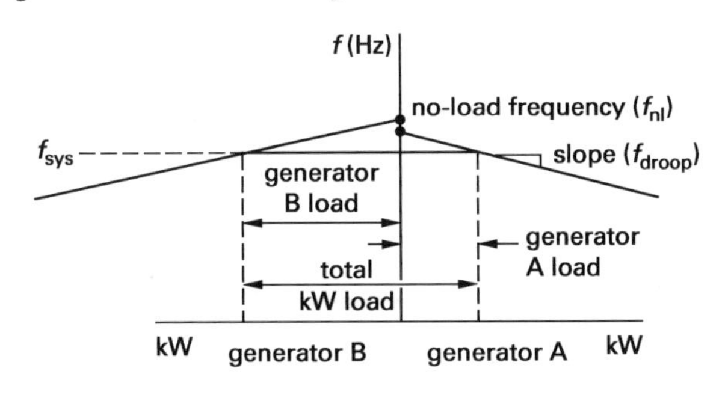

Figure 33.10 Reactive Load Sharing

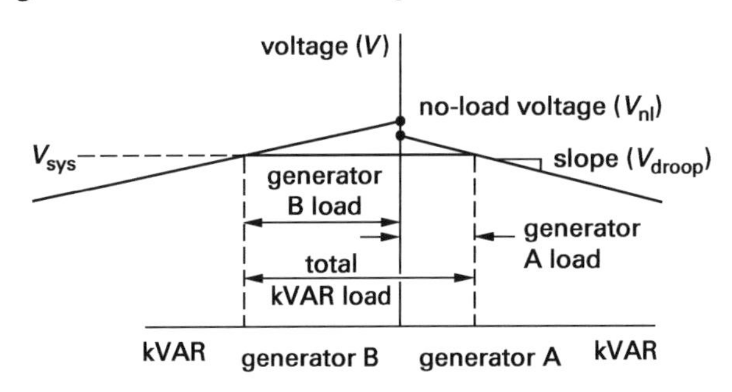

8. DIRECT CURRENT GENERATORS

Commutation is the process of current reversal in the armature windings that provides direct current to the brushes.

Armature reaction is the interaction between the magnetic flux produced by the armature current and the magnetic flux produced by the field current.

(See Fig. 33.13.)

EPRM Chapter 34
Three-Phase Electricity and Power

5. GENERATION OF THREE-PHASE POTENTIAL

(See Fig. 34.2.)

$$\mathbf{V}_a = V_p\angle 0° \quad 34.16$$

$$\mathbf{V}_b = V_p\angle{-120°} \quad 34.17$$

$$\mathbf{V}_c = V_p\angle{-240°} \quad 34.18$$

8. DELTA-CONNECTED LOADS

(See Fig. 34.5.)

$$|\mathbf{I}_A| = |\mathbf{I}_{AB} - \mathbf{I}_{CA}| = \sqrt{3}I_{AB} \quad 34.28$$

$$|\mathbf{I}_B| = |\mathbf{I}_{BC} - \mathbf{I}_{AB}| = \sqrt{3}I_{BC} \quad 34.29$$

$$|\mathbf{I}_C| = |\mathbf{I}_{CA} - \mathbf{I}_{BC}| = \sqrt{3}I_{CA} \quad 34.30$$

$$\begin{aligned} P_t &= 3P_p = 3V_pI_p\cos|\phi| \\ &= 3V_pI_p\text{pf} \\ &= \sqrt{3}VI\cos\phi \\ &= \sqrt{3}VI\text{pf} \end{aligned} \quad 34.32$$

Figure 33.13 *DC Generator Armature Flux*

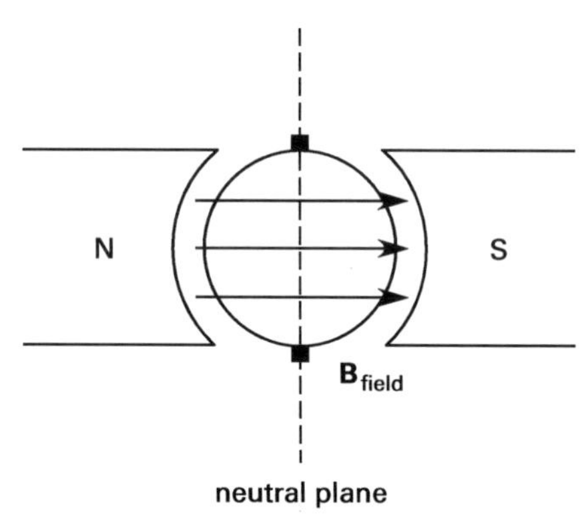

(a) stator magnetic field

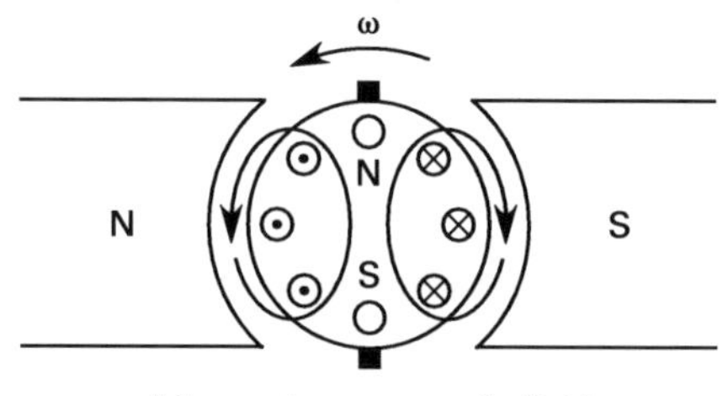

(b) armature magnetic field

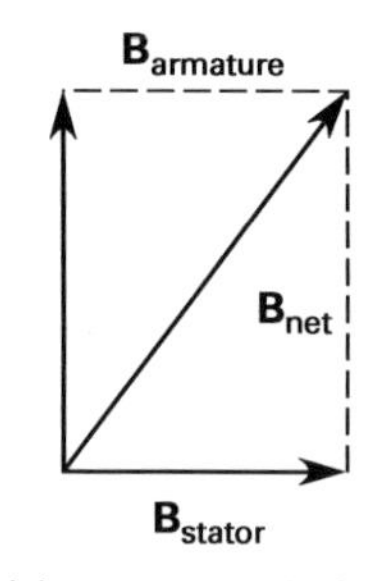

(c) net magnetic field

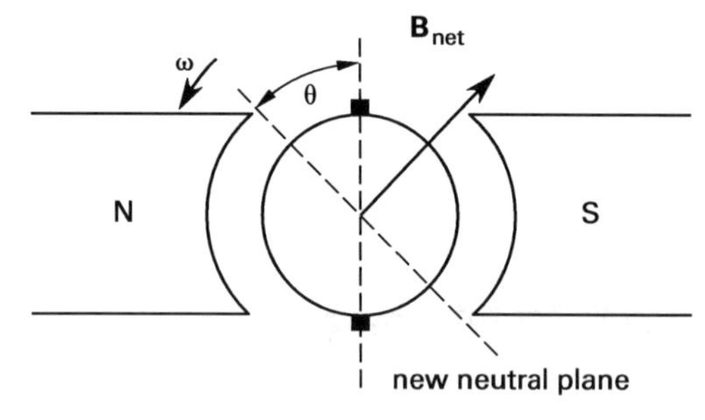

(d) neutral plane shift

Figure 34.2 *Three-Phase Voltage*

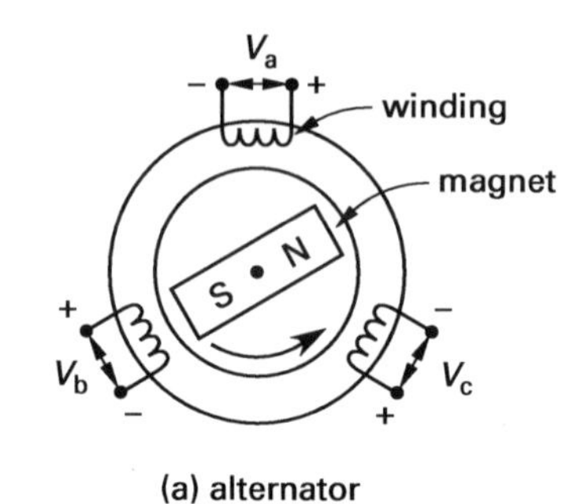

(a) alternator

V V_a V_b V_c

(b) ABC (positive) sequence

Figure 34.5 *Delta-Connected Loads*

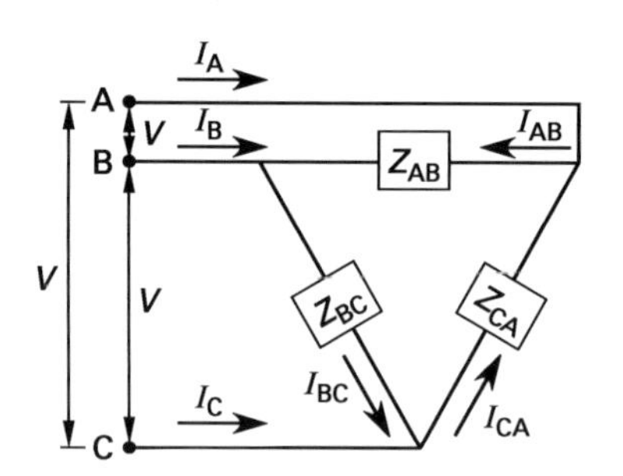

9. WYE-CONNECTED LOADS

$$\mathbf{I}_A = \mathbf{I}_{AN} = \frac{\mathbf{V}_{AN}}{\mathbf{Z}_{AN}} = \frac{\mathbf{V}_{AB}}{\sqrt{3}\mathbf{Z}_{AN}} \quad 34.33$$

$$\mathbf{I}_B = \mathbf{I}_{BN} = \frac{\mathbf{V}_{BN}}{\mathbf{Z}_{BN}} = \frac{\mathbf{V}_{BC}}{\sqrt{3}\mathbf{Z}_{BN}} \quad 34.34$$

$$\mathbf{I}_C = \mathbf{I}_{CN} = \frac{\mathbf{V}_{CN}}{\mathbf{Z}_{CN}} = \frac{\mathbf{V}_{CA}}{\sqrt{3}\mathbf{Z}_{CN}} \quad 34.35$$

$$\mathbf{I}_N = 0 \quad \text{[balanced]} \quad 34.36$$

$$P_t = 3P_p = 3V_pI_p\cos\phi = \sqrt{3}VI\cos\phi \quad 34.37$$

Figure 34.6 Wye-Connected Loads

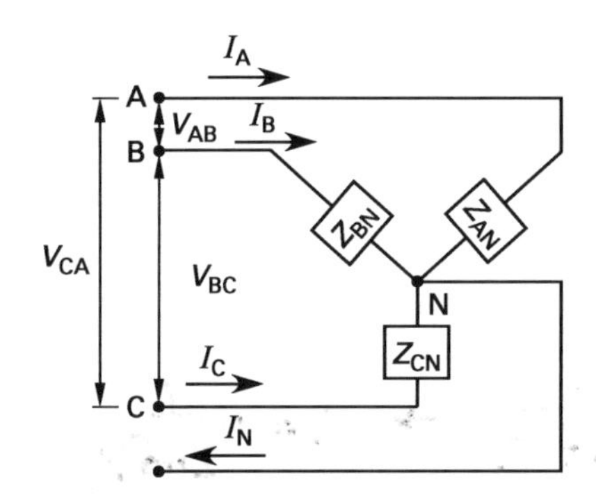

11. PER-UNIT CALCULATIONS

$$\text{per unit} = \frac{\text{actual}}{\text{base}} = \frac{\text{percent}}{100\%} \qquad 34.38$$

For a three-phase system, the usual bases are the line voltage (in V) and the total (three-phase) apparent power (in VA) ratings.

$$V_p = \frac{V_l}{\sqrt{3}} \qquad 34.39$$

$$S_p = \frac{S_t}{3} \qquad 34.40$$

The per-unit system is represented by Eq. 34.41 through Eq. 34.46. It is presented elsewhere as phase bases for use with single-phase systems. Conversion to a three-phase base, which uses line and total quantities, is accomplished with Eq. 34.39 and Eq. 34.40 and is indicated by the subscript 3ϕ in Eq. 34.41 through Eq. 34.46.

$$S_{\text{base}} = S_p = \left(\frac{S}{3}\right)_{3\phi} \qquad 34.41$$

$$V_{\text{base}} = V_p = \left(\frac{V}{\sqrt{3}}\right)_{3\phi} \qquad 34.42$$

$$I_{\text{base}} = \frac{S_{\text{base}}}{V_{\text{base}}} = \frac{S_p}{V_p} = \left(\frac{S}{\sqrt{3}V}\right)_{3\phi} \qquad 34.43$$

$$Z_{\text{base}} = \frac{V_{\text{base}}}{I_{\text{base}}} = \frac{V_p^2}{S_p} = \left(\frac{V^2}{S}\right)_{3\phi} = \left(\frac{V}{\sqrt{3}I}\right)_{3\phi} \qquad 34.44$$

$$P_{\text{base}} = P_p = \left(\frac{P}{3}\right)_{3\phi} \qquad 34.45$$

$$Q_{\text{base}} = Q_p = \left(\frac{Q}{3}\right)_{3\phi} \qquad 34.46$$

The per-unit values are

$$I_{\text{pu}} = \frac{I_{\text{actual}}}{I_{\text{base}}} \qquad 34.47$$

$$V_{\text{pu}} = \frac{V_{\text{actual}}}{V_{\text{base}}} \qquad 34.48$$

$$Z_{\text{pu}} = \frac{Z_{\text{actual}}}{Z_{\text{base}}} \qquad 34.49$$

$$P_{\text{pu}} = \frac{P_{\text{actual}}}{P_{\text{base}}} \qquad 34.50$$

$$Q_{\text{pu}} = \frac{Q_{\text{actual}}}{Q_{\text{base}}} \qquad 34.51$$

Ohm's law and other circuit analysis methods can be used with the per-unit quantities.

$$V_{\text{pu}} = I_{\text{pu}} Z_{\text{pu}} \qquad 34.52$$

The general method for converting from one per-unit value, call it χ, to another in a different base is

$$\chi_{\text{pu,new}} = \chi_{\text{pu,old}} \left(\frac{\chi_{\text{base,old}}}{\chi_{\text{base,new}}}\right) \qquad 34.53$$

The impedance per-unit value conversion is

$$Z_{\text{pu,new}} = Z_{\text{pu,old}} \left(\frac{V_{\text{base,old}}}{V_{\text{base,new}}}\right)^2 \left(\frac{S_{\text{base,new}}}{S_{\text{base,old}}}\right) \qquad 34.54$$

12. UNBALANCED LOADS

Unbalanced systems can be evaluated by computing the phase currents and then applying Kirchhoff's current law (in vector form) to obtain the line currents. The *neutral current* is

$$\mathbf{I}_N = -(\mathbf{I}_A + \mathbf{I}_B + \mathbf{I}_C) \qquad 34.55$$

Power Factor CORRection

Parallel (Cap): $Q_{phase} = \frac{V^2}{X_L}$

$X_c = -\frac{1}{\omega C} = -\frac{1}{2\pi f C} \rightarrow C = -\frac{1}{2\pi f X}$

Series (Ind): $Q_{phase} = \frac{I^2}{X_L}$

Distribution

EPRM Chapter 36
Power Distribution

6. UNDERGROUND DISTRIBUTION

- *optimal thickness*

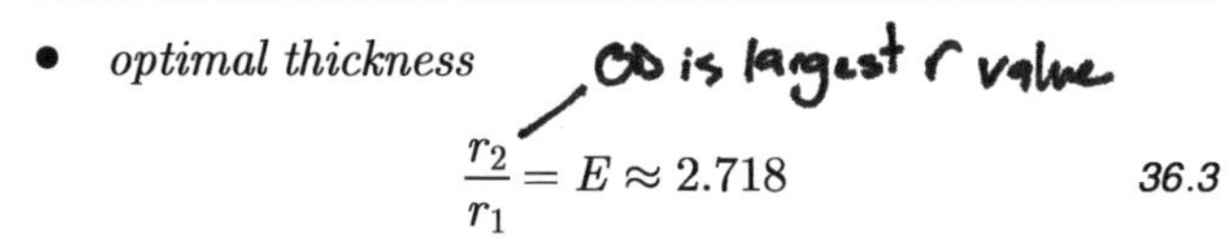

$$\frac{r_2}{r_1} = E \approx 2.718 \qquad 36.3$$

- *operating voltage for a capacitance-graded cable*

$$V = E_{\text{max}}\left(r_1 \ln \frac{r_2}{r_1} + r_2 \ln \frac{r_3}{r_2}\right) \qquad 36.4$$

7. FAULT ANALYSIS: SYMMETRICAL

A three-phase *symmetrical fault*, such as that in Fig. 36.7(d), has three specific time periods of concern, as shown in Fig. 36.8.

Figure 36.7 *Fault Types, from Most Likely to Least Likely*

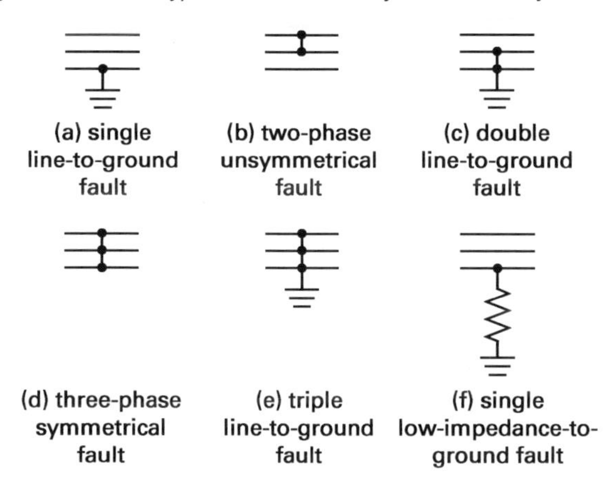

The voltage E''_g is calculated for the subtransient interval *just prior to the initiation of the fault* using

$$E''_g = V_t + jI_L X''_d \qquad 36.11$$

During the *transient period* the correct generator voltage for this period is calculated *just prior to the initiation of the fault* using

$$E'_g = V_t + jI_L X'_d \qquad 36.12$$

(See Fig. 36.9.)

Figure 36.8 *Symmetrical Fault Terminology*

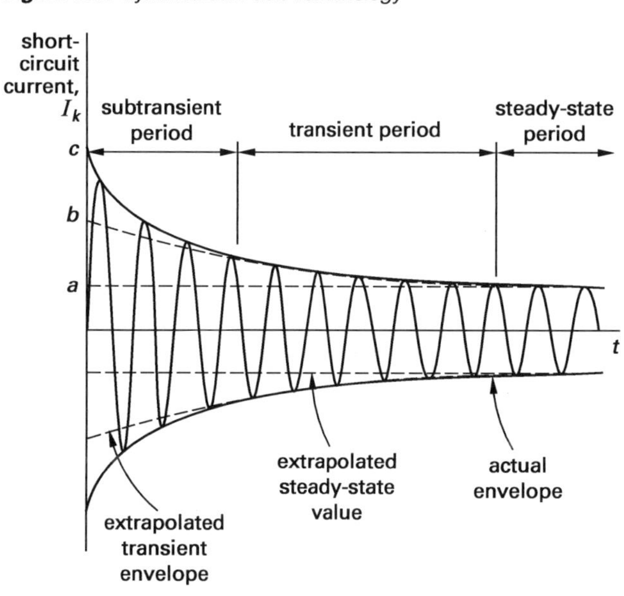

(a) synchronous generator three-phase fault response

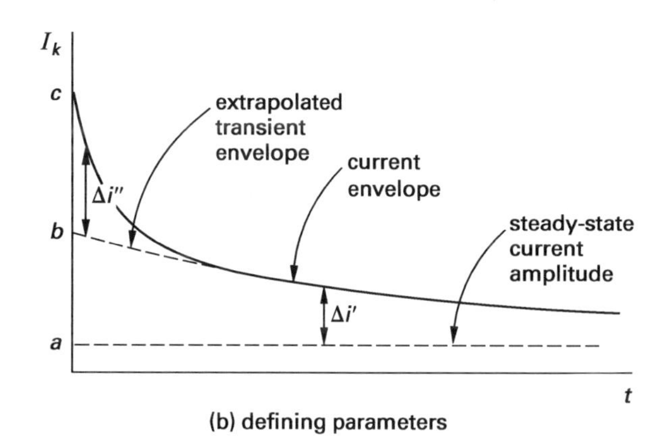

(b) defining parameters

8. FAULT ANALYSIS: UNSYMMETRICAL

Unsymmetrical faults, also called *asymmetrical faults*, are any faults other than a three-phase short.

The unsymmetrical phasors of Fig. 36.11 are represented in terms of their symmetrical components by the following equations.

$$V_A = V_{a0} + V_{a1} + V_{a2} \qquad 36.14$$

$$V_B = V_{b0} + V_{b1} + V_{b2} \qquad 36.15$$

$$V_C = V_{c0} + V_{c1} + V_{c2} \qquad 36.16$$

Figure 36.9 Synchronous Generator Fault Models

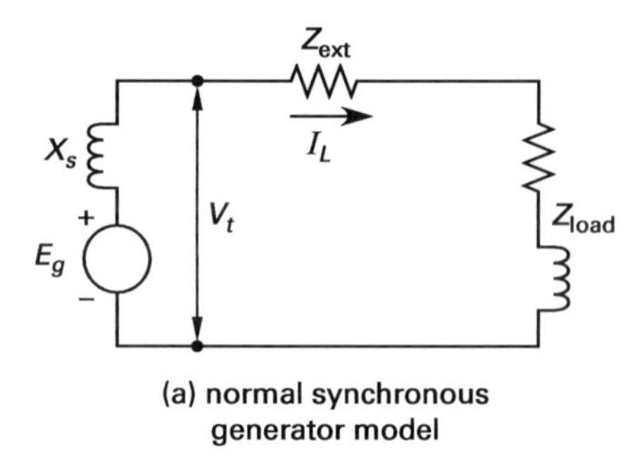

(a) normal synchronous generator model

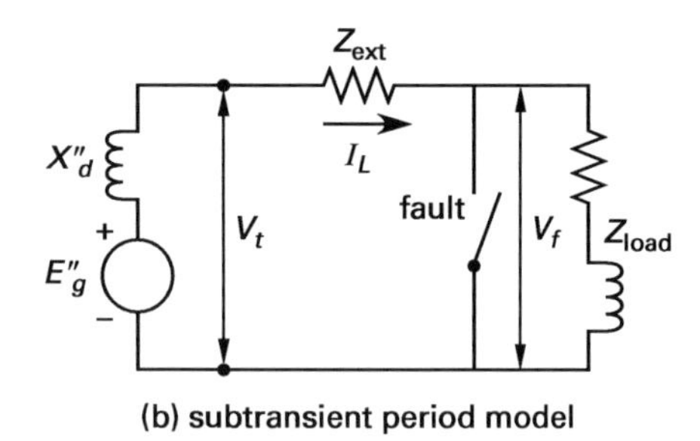

(b) subtransient period model

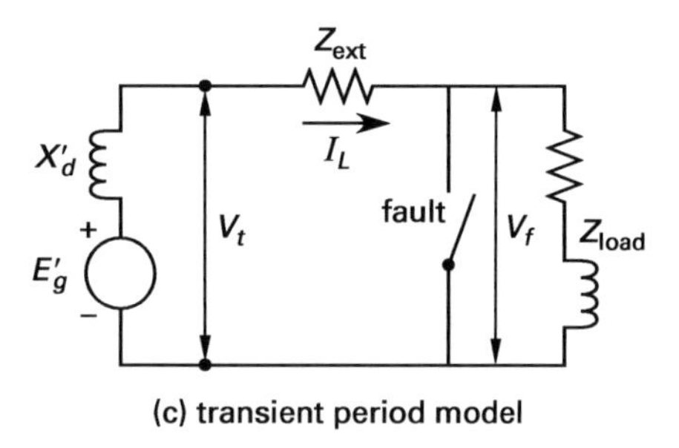

(c) transient period model

Figure 36.11 Phasor Diagram: Symmetrical Components of Unbalanced Phasors

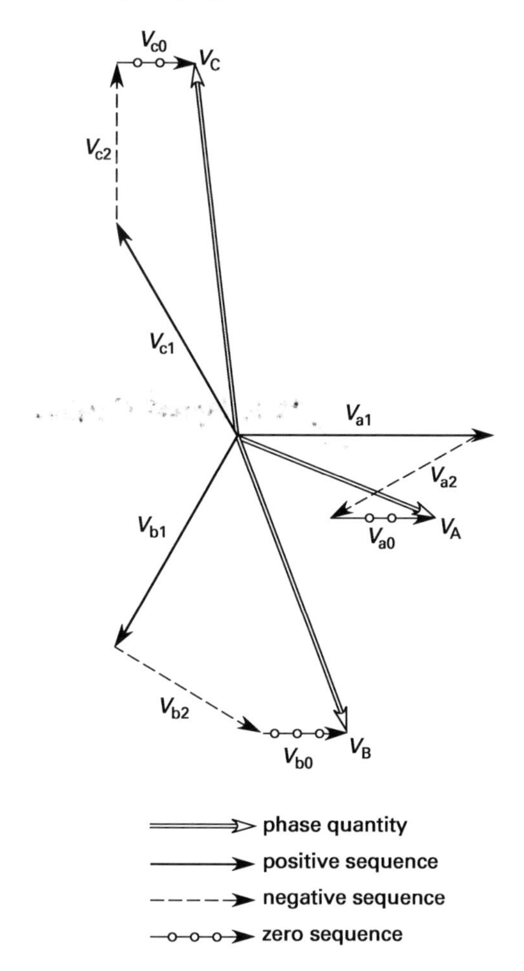

Consider the operator a defined as $1\angle 120°$, a unit vector with an angle of 120°.

$$a = 1\angle 120° = 1 \times e^{j120°}$$

$$= -0.5 + j0.866 \quad \text{36.17}$$

$$a^2 = 1\angle 240° = -0.5 - j0.866 = a^* \quad \text{36.18}$$

$$a^3 = 1\angle 360° = 1\angle 0° \quad \text{36.19}$$

$$a^4 = a \quad \text{36.20}$$

$$a^5 = a^2 \quad \text{36.21}$$

$$a^6 = a^3 \quad \text{36.22}$$

$$1 + a + a^2 = 0 \quad \text{36.23}$$

The unsymmetrical components can be represented in terms of a single phase. (See Fig. 36.12.)

$$V_A = V_{a0} + V_{a1} + V_{a2} \quad \text{36.24}$$

$$V_B = V_{a0} + a^2 V_{a1} + a V_{a2} \quad \text{36.25}$$

$$V_C = V_{a0} + a V_{a1} + a^2 V_{a2} \quad \text{36.26}$$

Solving for the sequence components,

$$V_{a0} = \tfrac{1}{3}(V_A + V_B + V_C) \quad \text{36.27}$$

$$V_{a1} = \tfrac{1}{3}(V_A + a V_B + a^2 V_C) \quad \text{36.28}$$

$$V_{a2} = \tfrac{1}{3}(V_A + a^2 V_B + a V_C) \quad \text{36.29}$$

EPRM Chapter 37 **Power Transformers**

1. THEORY

Figure 37.1 Transformer Model

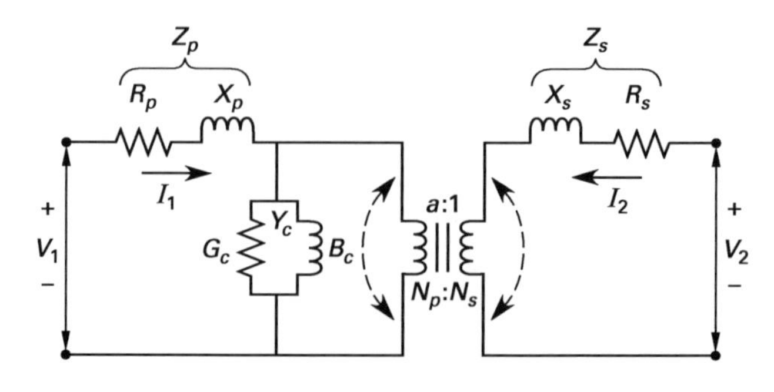

Figure 36.12 Components of Unsymmetrical Phasors

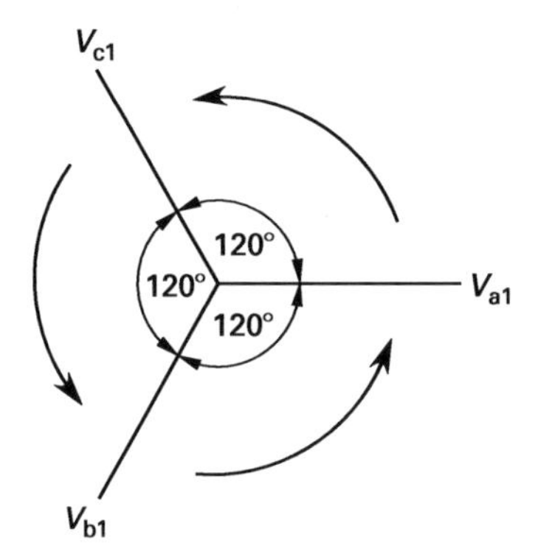

(a) positive sequence

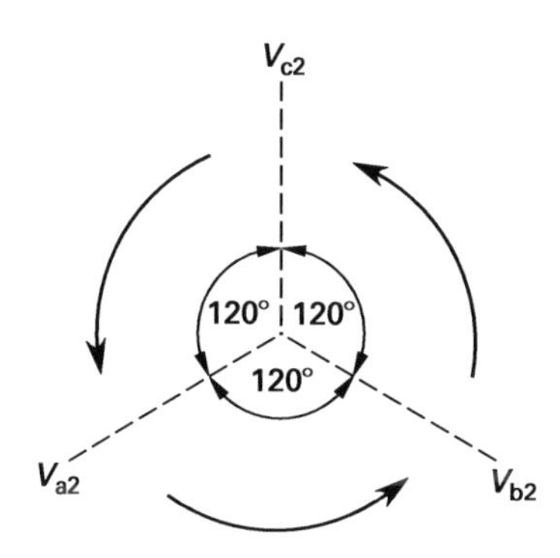

(b) negative sequence

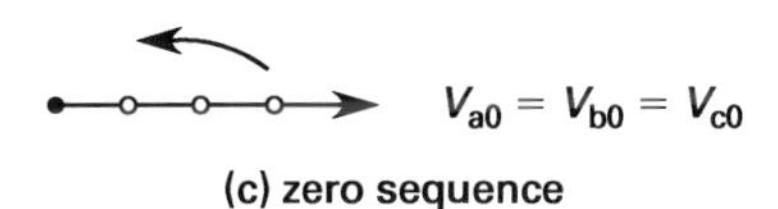

(c) zero sequence

2. TRANSFORMER RATING

- *eddy current losses*, P_e
- *hysteresis losses*, P_h
- *mass of iron*, m
- *maximum flux density*, B_{max}
- *Steinmetz exponent*, the exponent n
- *coupling coefficient*, k

$$P_e = k_e B_{\text{max}}^2 f^2 m \qquad 37.1$$

$$P_h = k_h B_{\text{max}}^n f m \qquad 37.2$$

- *core losses*

$$P_c = \frac{V_1^2}{R_c} \qquad 37.3$$

- *copper losses*

$$P_{\text{Cu}} = I^2 R = I_1^2 R_p + I_2^2 R_s \qquad 37.4$$

- *transformer efficiency*

$$\eta = \frac{P_{\text{out}}}{P_{\text{in}}} = \frac{P_{\text{in}} - \sum P_{\text{losses}}}{P_{\text{in}}} = \frac{P_{\text{out}}}{P_{\text{out}} + P_c + P_{\text{Cu}}} \qquad 37.5$$

4. VOLTAGE REGULATION

$$\text{VR} = \frac{V_{\text{nl}} - V_{\text{fl}}}{V_{\text{fl}}} \qquad 37.6$$

$$\text{VR} = \frac{\frac{V_p}{a} - V_{s,\text{rated}}}{V_{s,\text{rated}}} \qquad 37.7$$

5. CONNECTIONS

$$V_l = V_\phi \quad \text{[delta]} \qquad 37.8$$

$$I_l = \sqrt{3} I_\phi \quad \text{[delta]} \qquad 37.9$$

$$V_l = \sqrt{3} V_\phi \quad \text{[wye]} \qquad 37.10$$

$$I_l = I_\phi \quad \text{[wye]} \qquad 37.11$$

$$P = \sqrt{3} I_l V_l (\text{pf}) = 3 I_\phi V_\phi (\text{pf}) \qquad 37.12$$

7. OPEN-CIRCUIT TEST

The open-circuit test determines the core parameters and the turns ratio.

Figure 37.4 Transformer Open-Circuit Test Model

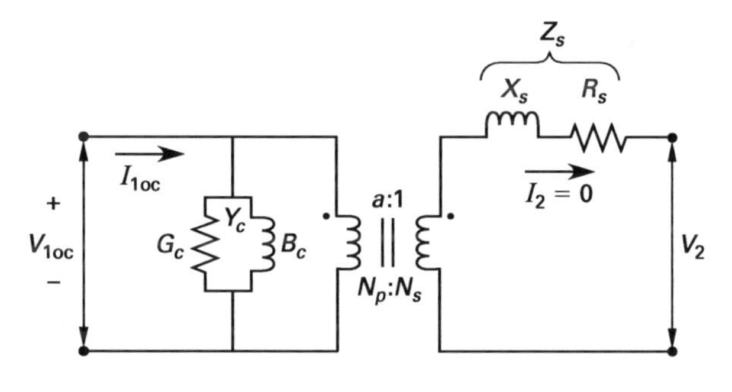

- *admittance*

$$Y_c = G_c + jB_c = \frac{I_{1\text{oc}}}{V_{1\text{oc}}} \qquad 37.13$$

- *conductance*

$$G_c = \frac{P_{\text{oc}}}{V_{1\text{oc}}^2} \qquad 37.14$$

- *susceptance*

$$B_c = \frac{1}{X_c} = \frac{-1}{\omega L_c} = -\sqrt{Y_c^2 - G_c^2} = \frac{-\sqrt{I_{1\text{oc}}^2 V_{1\text{oc}}^2 - P_{\text{oc}}^2}}{V_{1\text{oc}}^2} \qquad 37.15$$

- *turns ratio*

$$a_{ps} = \frac{V_{1oc}}{V_{2oc}} \quad 37.16$$

- *power*

$$P_{oc} = V_{1oc}^2 G_c \quad 37.17$$

- *reactive power*

$$Q_{oc} = V_{1oc}^2 B_c \quad 37.18$$

- *apparent power*

$$S_{oc} = V_{1oc}^2 Y_c = V_{1oc} I_{1oc} = \sqrt{P_{oc}^2 + Q_{oc}^2} \quad 37.19$$

8. SHORT-CIRCUIT TEST

The short-circuit test determines the winding impedances and verifies the turns ratio.

$$Z = R_p + jX_p + a_{ps}^2 (R_s + jX_s) = \frac{V_{1sc}}{I_{1sc}} \quad 37.20$$

Figure 37.5 *Transformer Short-Circuit Test Model*

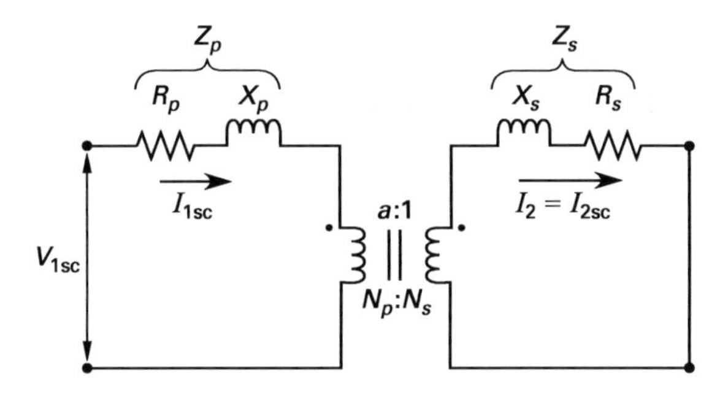

- *total resistance, R*

$$R = R_p + a_{ps}^2 R_s = \frac{P_{sc}}{I_{1sc}^2} \quad 37.21$$

To maximize efficiency, transformers are normally designed with R_p equal to $a_{ps}^2 R_s$.

$$R_p = a_{ps}^2 R_s = \frac{P_{sc}}{2I_{1sc}^2} \quad 37.22$$

- *total reactance, X*

$$X = X_p + a_{ps}^2 X_s = \frac{\sqrt{I_{sc}^2 V_{sc}^2 - P_{sc}^2}}{I_{1sc}^2} \quad 37.23$$

To maximize efficiency, transformers are normally designed with X_p equal to $a_{ps}^2 X_s$.

$$X_p = a_{ps}^2 X_s = \frac{Q_{sc}}{2I_{1sc}^2} \quad 37.24$$

- *turns ratio*

$$a_{ps} = \frac{I_{2sc}}{I_{1sc}} \quad 37.25$$

- *power*

$$P_{sc} = I_{1sc}^2 R = I_{1sc}^2 (R_p + a_{ps}^2 R_s) \quad 37.26$$

- *reactive power*

$$Q_{sc} = I_{1sc}^2 X = I_{1sc}^2 (X_p + a_{ps}^2 X_s) \quad 37.27$$

- *apparent power*

$$S_{sc} = I_{1sc}^2 Z_{sc} = V_{1sc} I_{1sc} = \sqrt{P_{sc}^2 + Q_{sc}^2} \quad 37.28$$

9. ABCD PARAMETERS

- *ABCD parameters for any two-port network*

$$V_{in} = \mathrm{A} V_{out} - \mathrm{B} I_{out} \quad 37.29$$

$$I_{in} = \mathrm{C} V_{out} - \mathrm{D} I_{out} \quad 37.30$$

EPRM Chapter 38
Transmission Lines

1. FUNDAMENTALS

$$Z_0 = \sqrt{\frac{Z_l}{Y_l}} \quad [\text{in } \Omega] \quad 38.1$$

$$\Gamma = \frac{V_{\text{reflected}}}{V_{\text{incident}}} = \frac{I_{\text{reflected}}}{I_{\text{incident}}} = \left|\frac{Z_{\text{load}} - Z_0}{Z_{\text{load}} + Z_0}\right| \quad 38.3$$

The fraction of incident power that is reflected back to the source from the load is Γ^2.

$$\Gamma = \frac{\text{SWR} - 1}{\text{SWR} + 1} \quad 38.4$$

- *velocity of propagation*

$$\text{v}_w = \frac{1}{\sqrt{L_l C_l}} \quad 38.5$$

3. SKIN EFFECT

$$\delta = \frac{1}{\sqrt{\frac{\pi f \mu}{\rho}}} = \frac{1}{\sqrt{\pi f \mu \sigma}} \quad 38.8$$

- *AC resistance per unit length for a flat conducting plate of unit width w*

$$R_{l,\text{AC}} = \frac{\rho}{\delta w} \quad [\text{in } \Omega/\text{m}] \quad 38.9$$

$$\delta_{\text{Cu}} = \frac{0.066}{\sqrt{f}} \quad [\text{in m}] \quad 38.10$$

7. SINGLE-PHASE INDUCTANCE

The total inductance of a single-phase system, which consists of two conductors, is the sum of the internal and external inductances.

$$
\begin{aligned}
L_l &= L_{l,\text{int}} + L_{l,\text{ext}} \\
&= \left(\frac{\mu_0}{4\pi}\right)\left(1 + 4\ln\frac{D}{r}\right) \quad \text{[in H/m]}
\end{aligned} \qquad 38.19
$$

Equation 38.19 is simplified using the concept of the *geometric mean radius* (GMR).

$$
\text{GMR} = re^{-1/4} \qquad 38.20
$$

$$
\begin{aligned}
L_l &= \frac{\mu_0}{\pi}\ln\frac{D}{\text{GMR}} \\
&= (4\times 10^{-7})\ln\frac{D}{\text{GMR}} \quad \text{[in H/m]}
\end{aligned} \qquad 38.21
$$

8. SINGLE-PHASE CAPACITANCE

$$
C_l = \frac{\pi\epsilon_0}{\ln\frac{D}{r}} = \frac{2.78\times 10^{-11}}{\ln\frac{D}{r}} \quad \text{[in F/m]} \qquad 38.22
$$

9. THREE-PHASE TRANSMISSION

When the conductors are not symmetrically arranged, as is often the case, formulas used for the inductance and capacitance are still valid if the equivalent distance, D_e, is substituted for the distance.

$$
D_e = \sqrt[3]{D_{\text{ab}}D_{\text{bc}}D_{\text{ca}}} \qquad 38.23
$$

- *per-phase inductance per unit length of a three-phase transmission line*

$$
\begin{aligned}
L_l &= \frac{\mu_0}{2\pi}\ln\frac{D_e}{\text{GMR}} \\
&= (2\times 10^{-7})\ln\frac{D_e}{\text{GMR}} \quad \text{[in H/m]}
\end{aligned} \qquad 38.24
$$

- *per-phase capacitance per unit length of a three-phase transmission line*

$$
C_l = \frac{2\pi\epsilon_0}{\ln\frac{D_e}{r}} = \frac{5.56\times 10^{-11}}{\ln\frac{D_e}{r}} \quad \text{[in F/m]} \qquad 38.25
$$

10. POWER TRANSMISSION LINES

$$
\mathbf{V}_l = \mathbf{I}R_l + j\mathbf{I}X_l = \mathbf{I}\mathbf{Z}_l \qquad 38.26
$$

$$
X_L = (2.022\times 10^{-3})f\ln\frac{D}{\text{GMR}} \quad [\text{in }\Omega/\text{mi}] \qquad 38.27
$$

For line-to-line spacings other than one foot, the correction factor given by Eq. 38.28 must be applied.

$$
K_L = 1 + \frac{\ln D}{\ln\frac{1}{\text{GMR}}} \qquad 38.28
$$

- *capacitive reactance per mile* (approximate)

$$
X_C = \frac{1.781}{f}\ln\frac{D}{r} \quad [\text{in M}\Omega\text{-mi}] \qquad 38.29
$$

- *correction factor* (for line-to-line spacings other than one foot)

$$
K_C = 1 + \frac{\ln D}{\ln\frac{1}{r}} \qquad 38.30
$$

11. TRANSMISSION LINE REPRESENTATION

$$
\text{VR} = \frac{|V_{R,\text{nl}}| - |V_{R,\text{fl}}|}{|V_{R,\text{fl}}|} \qquad 38.31
$$

$$
\eta_P = \frac{P_R}{P_S} \qquad 38.32
$$

(See Table 38.3.)

Figure 38.3 *Transmission Line Two-Port Network*

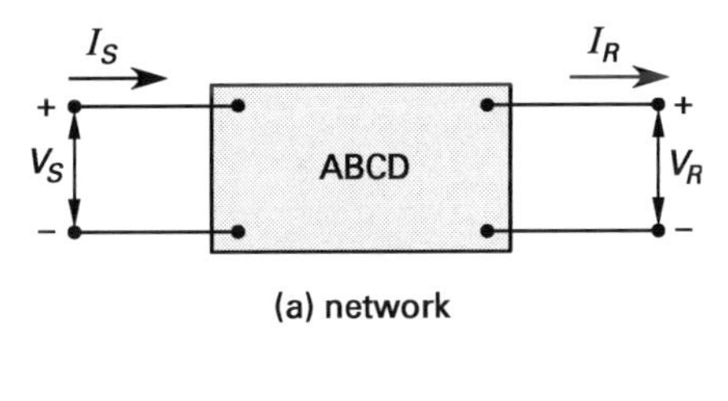

(a) network

$V_S = AV_R + BI_R$

$I_S = CV_R + DI_R$

(b) equations

2×2

$$
\begin{bmatrix} V_S \\ I_S \end{bmatrix} = \begin{bmatrix} A & B \\ C & D \end{bmatrix}\begin{bmatrix} V_R \\ I_R \end{bmatrix}
$$

(c) matrix form of equations

12. SHORT TRANSMISSION LINES

Short transmission lines are 60 Hz lines that are less than 80 km (50 mi) long.

Figure 38.4 *Short Transmission Line Model*

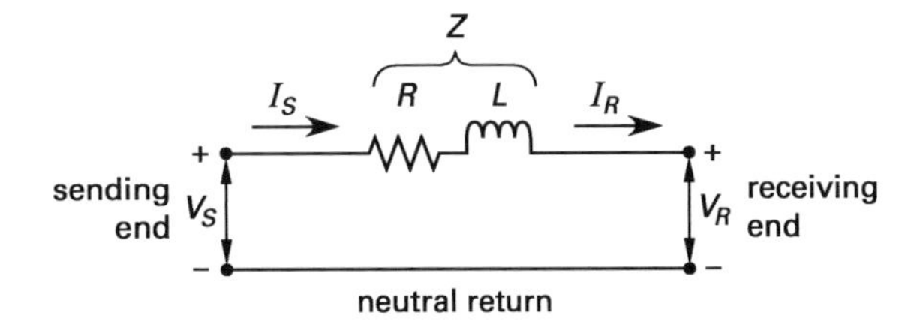

Table 38.3 *Per-Phase ABCD Constants for Transmission Lines*

transmission line length	equivalent circuit	A	B	C	D
short <80 km (50 mi)	series impedance Fig. 38.4	1	Z	0	1
medium 80–240 km (50–150 mi)	nominal-T Fig. 38.5(a)	$1+\frac{1}{2}YZ$	$Z(1+\frac{1}{4}YZ)$	Y	$1+\frac{1}{2}YZ$
medium 80–240 km (50–150 mi)	nominal-π Fig. 38.5(b)	$1+\frac{1}{2}YZ$	Z	$Y(1+\frac{1}{4}YZ)$	$1+\frac{1}{2}YZ$
long >240 km (150 mi)	distributed parameters Fig. 38.6	$\cosh \gamma l$	$Z_0 \sinh \gamma l$	$\dfrac{\sinh \gamma l}{Z_0}$	$\cosh \gamma l$

13. MEDIUM-LENGTH TRANSMISSION LINES

Medium-length transmission lines are 60 Hz lines between 80 km and 240 km (50 mi and 150 mi) long.

Figure 38.5 *Medium-Length Transmission Line Models*

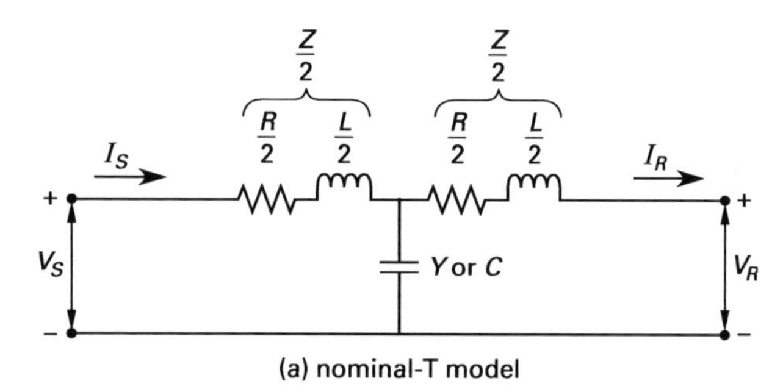

(a) nominal-T model

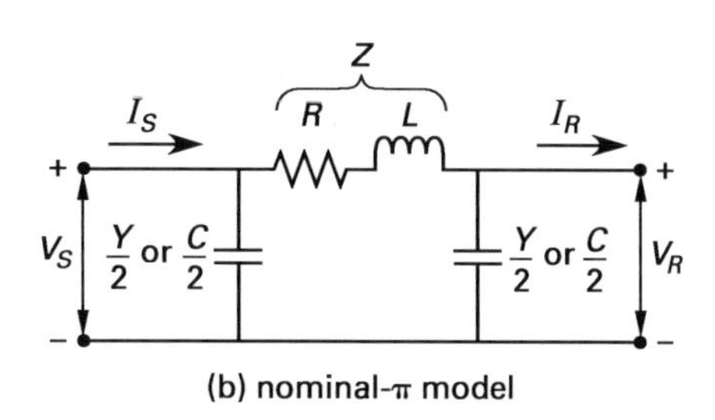

(b) nominal-π model

14. LONG TRANSMISSION LINES

Long transmission lines are 60 Hz lines greater than 240 km (150 mi) long.

$$Y_l = jB_{C,l} = \frac{j}{X_{C,l}} = -\frac{1}{jX_{C,l}} = j\omega C_l \qquad 38.43$$

(See Fig. 38.6.)

15. REFLECTION COEFFICIENT

$$\text{VSWR} = \frac{V_{\text{max}}}{V_{\text{min}}} \qquad 38.56$$

Figure 38.6 *Long Transmission Line Model*

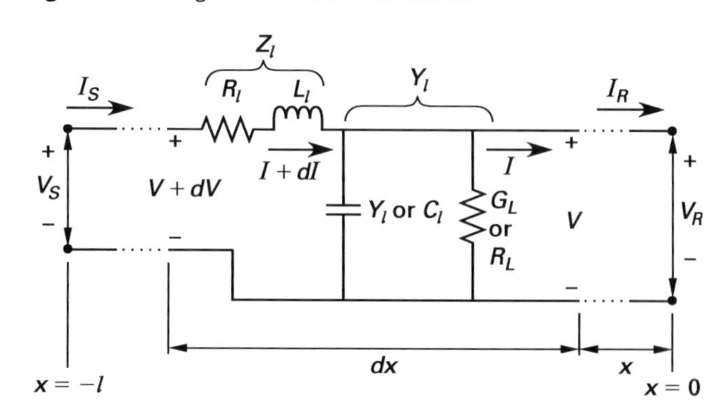

$$\text{ISWR} = \frac{I_{\text{max}}}{I_{\text{min}}} \qquad 38.57$$

$$\Gamma_L = \frac{V_{\text{reflected}}}{V_{\text{incident}}} = \frac{Z_{\text{load}} - Z_0}{Z_{\text{load}} + Z_0} \qquad 38.58$$

$$\Gamma_L = \frac{I_{\text{reflected}}}{I_{\text{incident}}} = \frac{Z_0 - Z_{\text{load}}}{Z_0 + Z_{\text{load}}} \qquad 38.59$$

The fraction of incident power that is reflected back to the source from the load is Γ^2.

$$\Gamma = \frac{\text{SWR} - 1}{\text{SWR} + 1} \qquad 38.60$$

17. HIGH-FREQUENCY TRANSMISSION LINES

At high frequencies, approximately 1 MHz and higher, wavelengths are shorter and even a few feet of line are treated as a long transmission line.

$$Z_l = jX_{L,l} = j\omega L_l \qquad 38.68$$

- *propagation constant*, γ (in units of radians per meter (mile)). β is the phase constant.

$$\gamma = \sqrt{Y_l Z_l} = \sqrt{(j\omega C_l)(j\omega L_l)} = j\omega\sqrt{L_l C_l} = j\beta \qquad 38.69$$

- *phase velocity*

$$\mathrm{v}_{\text{phase}} = \frac{1}{\sqrt{L_l C_l}} \qquad 38.70$$

- *wavelength*

$$\lambda = \frac{2\pi}{\beta} = \frac{1}{f\sqrt{L_l C_l}} \qquad 38.71$$

- *characteristic impedance*

$$Z_0 = \sqrt{\frac{Z_l}{Y_l}} = \sqrt{\frac{j\omega L_l}{j\omega C_l}} = \sqrt{\frac{L_l}{C_l}} \qquad 38.72$$

$$Z_{\text{in}} = Z_0\left(\frac{(1+\Gamma)\cos\beta l + j(1-\Gamma)\sin\beta l}{(1-\Gamma)\cos\beta l + j(1+\Gamma)\sin\beta l}\right) \qquad 38.77$$

$$Z_{\text{in}} = Z_0\left(\frac{Z_{\text{load}}\cos\beta l + jZ_0\sin\beta l}{Z_0\cos\beta l + jZ_{\text{load}}\sin\beta l}\right) \qquad 38.78$$

$$\text{VSWR} = \frac{|V^+| + |V^-|}{|V^+| - |V^-|} = \frac{1+|\Gamma|}{1-|\Gamma|} \qquad 38.79$$

$$Z_{\max} = Z_0(\text{VSWR}) \qquad 38.80$$

$$Z_{\min} = \frac{Z_0}{\text{VSWR}} \qquad 38.81$$

EPRM Chapter 39

The Smith Chart

1. FUNDAMENTALS

A *Smith chart* is a special polar diagram with constant reflection coefficient circles, constant standing wave ratio circles, constant resistance circles, constant reactance arcs (portions of a circle), and radius lines representing constant line-angle loci. The Smith chart is essentially a polar representation of the reflection coefficient in terms of the normalized resistance and reactance. The normalization occurs with respect to the characteristic impedance, Z_0. (See Fig. 39.1.)

$$z = \frac{Z}{Z_0} = \frac{R+jX}{Z_0} = \frac{1+\Gamma}{1-\Gamma} \qquad 39.1$$

Table 39.1 *Smith Chart Electrical Conditions*

electrical condition	reflection coefficient (Γ)	normalized resistance (r)	normalized reactance (x)
open circuit	$1\angle 0°$	∞ (arbitrary)	arbitrary (∞)
short circuit	$1\angle 180°$	0	0
pure reactance	$1\angle \pm 90°$	0	± 1
matched line (pure resistance)	0	1	0

Figure 39.1 *Smith Chart Components*

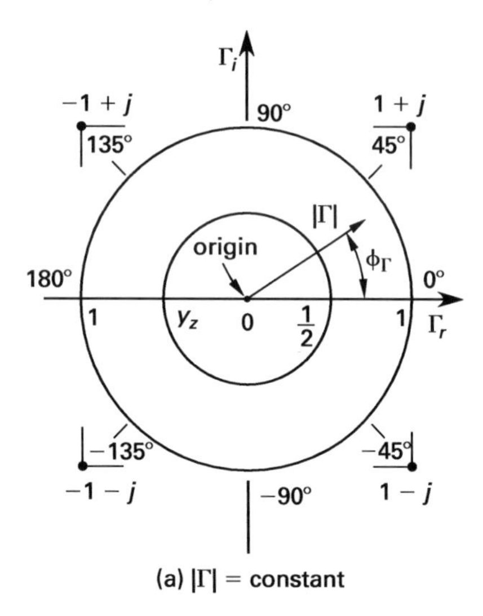

(a) |Γ| = constant

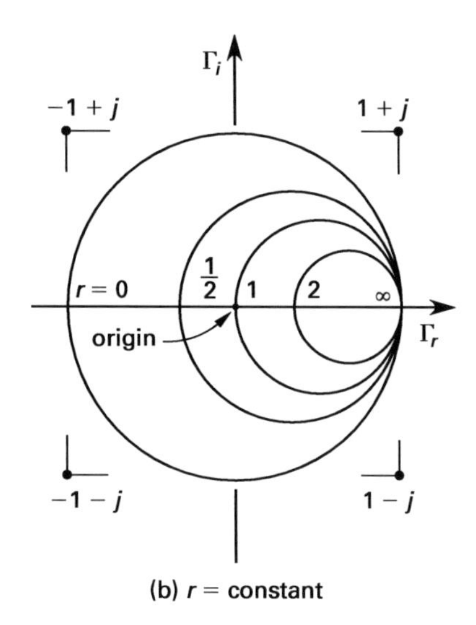

(b) r = constant

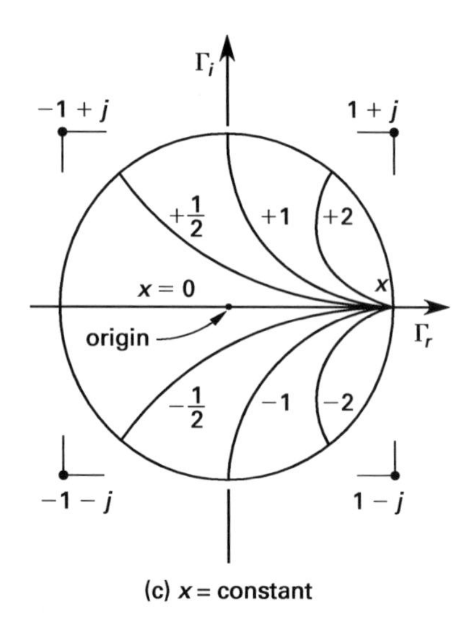

(c) x = constant

System Analysis

EPRM Chapter 40
Power System Analysis

2. POWER FLOW

$$\mathbf{S} = \mathbf{VI}^* = P + jQ = VI\cos\theta + jVI\sin\theta \qquad 40.1$$

3. THREE-PHASE CONNECTIONS

$$Z_{\mathrm{Y}} = \tfrac{1}{3}Z_{\Delta} \qquad 40.2$$

Figure 40.3 *120° Unity Phasor Addition*

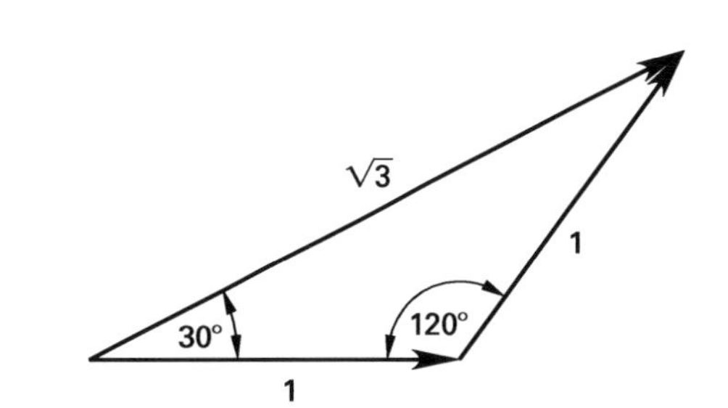

Figure 40.4 *Delta-Wye Transformations**

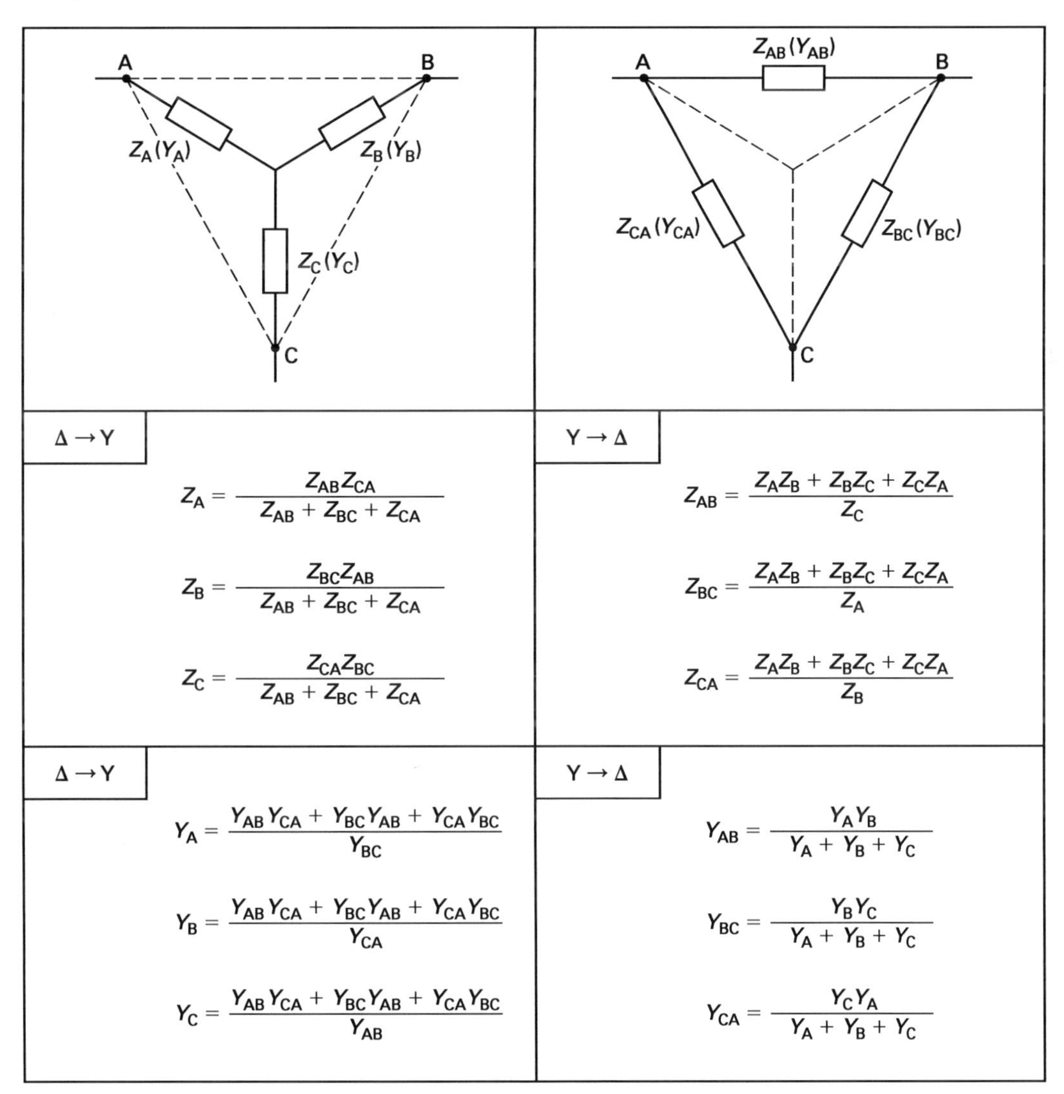

*Admittances and impedances with the same subscripts are reciprocals of one another.

$$P = 3V_pI_p\cos\theta = 3V_pI_p\text{pf} \qquad 40.3$$

$$P = \sqrt{3}V_lI_l\cos\theta = \sqrt{3}VI\text{pf} \qquad 40.4$$

4. OPERATOR: 120°

$$a = 1\angle 120° = 1e^{j2\pi/3} = -0.5 + j0.866 \qquad 40.7$$

$$a^2 = 1\angle 240° = 1e^{j4\pi/3} = -0.5 - j0.866 \qquad 40.8$$

$$a^3 = 1\angle 360° = 1e^{j2\pi} = 1\angle 0° = 1 \qquad 40.9$$

Figure 40.5 *The 120° Operator*

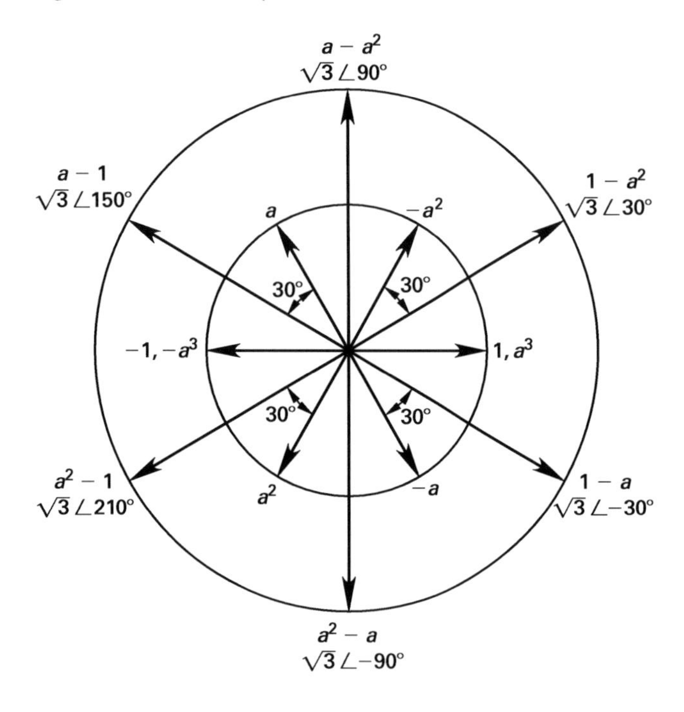

5. BALANCED THREE-PHASE CIRCUIT

(See Fig. 40.6.)

6. PER-UNIT SYSTEM

(See Table 40.1.)

7. SEQUENCE COMPONENTS

(See Fig. 40.8 and Fig. 40.9.)

EPRM Chapter 41 Analysis of Control Systems

1. TYPES OF RESPONSE

Natural response (also referred to as *initial condition response*, *homogeneous response*, and *unforced response*) is the manner in which a system behaves when energy is applied and then subsequently removed.

Forced response is the behavior of a system that is acted upon by a force that is applied periodically.

Figure 40.6 *Balanced Wye Voltage Relationships*

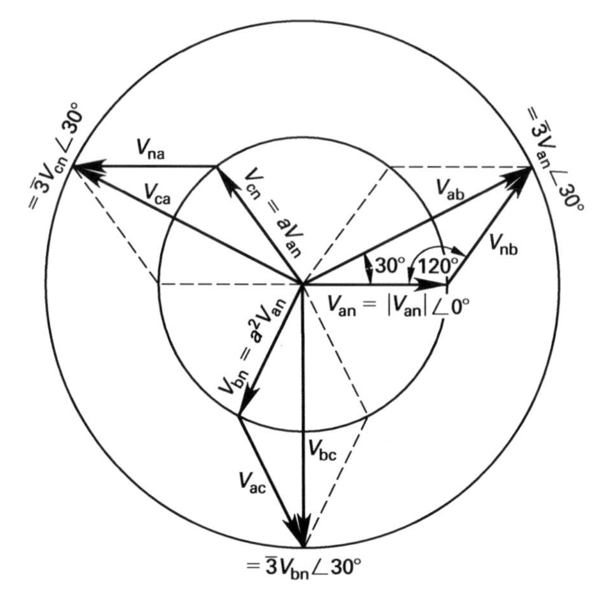

(a) line-to-line versus line-to-neutral voltages

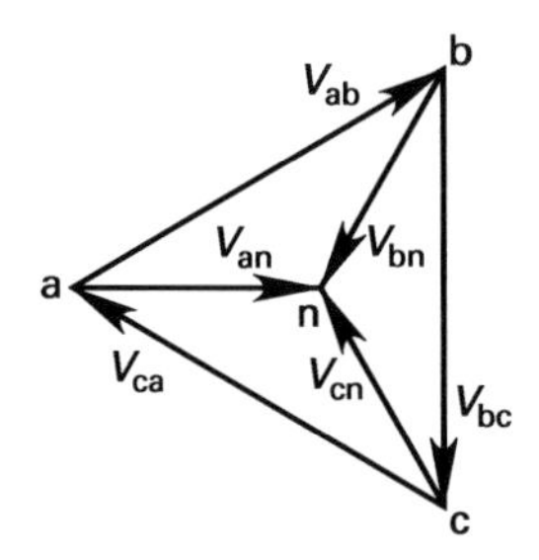

(b) alternative line-to-line versus line-to-neutral voltages

2. GRAPHICAL SOLUTION

When the system equation is a homogeneous second-order linear differential equation with constant coefficients in the form of Eq. 41.2, the natural time response can be determined from Fig. 41.1.

$$x'' + 2\zeta\omega x' + \omega^2 x = 0 \qquad 41.2$$

When the system equation is a second-order linear differential equation with constant coefficients and the forcing function is a step of height h (as in Eq. 41.3), the time response can be determined from Fig. 41.2.

$$x'' + 2\zeta\omega x' + \omega^2 x = \omega^2 h \qquad 41.3$$

- *damped frequency*

$$\omega_d = \omega\sqrt{1-\zeta^2} \qquad 41.4$$

Table 40.1 *Per-Unit Equations*[a,b]

$\text{per unit} = \frac{\text{actual}}{\text{base}} = \frac{\text{percent}}{100\%}$	$Z_{\text{base}} = \frac{V_{\text{base}}}{I_{\text{base}}} = \frac{V_p^2}{S_p} = \left(\frac{V^2}{S}\right)_{3\phi} = \left(\frac{V}{\sqrt{3}I}\right)_{3\phi}$
$S_{\text{base}} = S_p = \left(\frac{S}{3}\right)_{3\phi}$	$P_{\text{base}} = P_p = \left(\frac{P}{3}\right)_{3\phi}$
$V_{\text{base}} = V_p = \left(\frac{V}{\sqrt{3}}\right)_{3\phi}$	$Q_{\text{base}} = Q_p = \left(\frac{Q}{3}\right)_{3\phi}$
$I_{\text{base}} = \frac{S_{\text{base}}}{V_{\text{base}}} = \frac{S_p}{V_p} = \left(\frac{S}{\sqrt{3}V}\right)_{3\phi}$	$\chi_{\text{pu,new}} = \chi_{\text{pu,old}}\left(\frac{\chi_{\text{base,old}}}{\chi_{\text{base,new}}}\right)$
$Z_{\text{pu,new}} = Z_{\text{pu,old}}\left(\frac{V_{\text{base,old}}}{V_{\text{base,new}}}\right)^2\left(\frac{S_{\text{base,new}}}{S_{\text{base,old}}}\right)$	

[a]The equation for $\chi_{\text{pu,new}}$ can only be used to find current, power, or voltage.
[b]The equation for $Z_{\text{pu,new}}$ can only be used to find impedance.

Figure 40.8 *Symmetrical Components of Unbalanced Phasors*

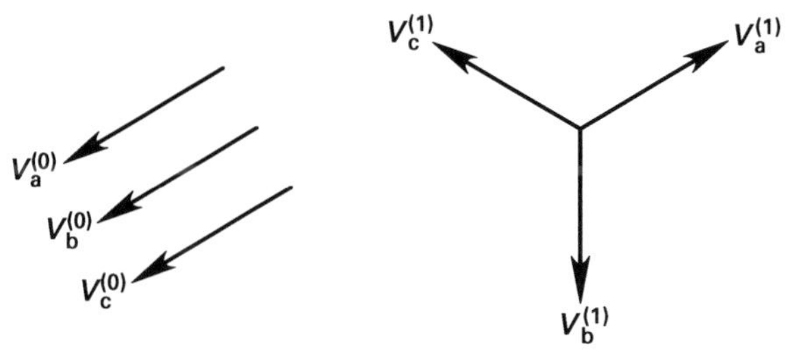

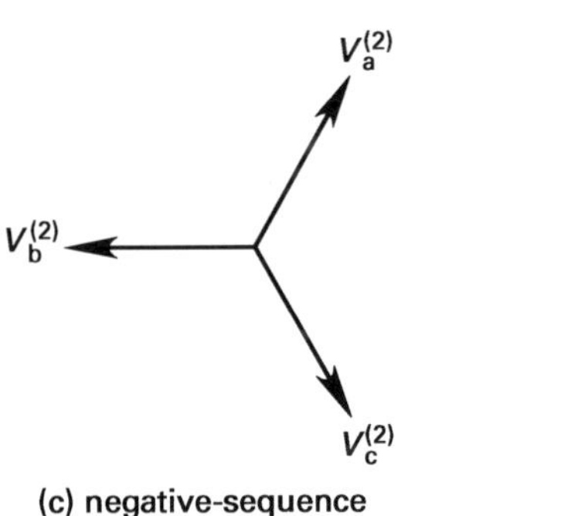

Figure 40.9 *Example of Unbalanced Phasor from Symmetrical Phasors*

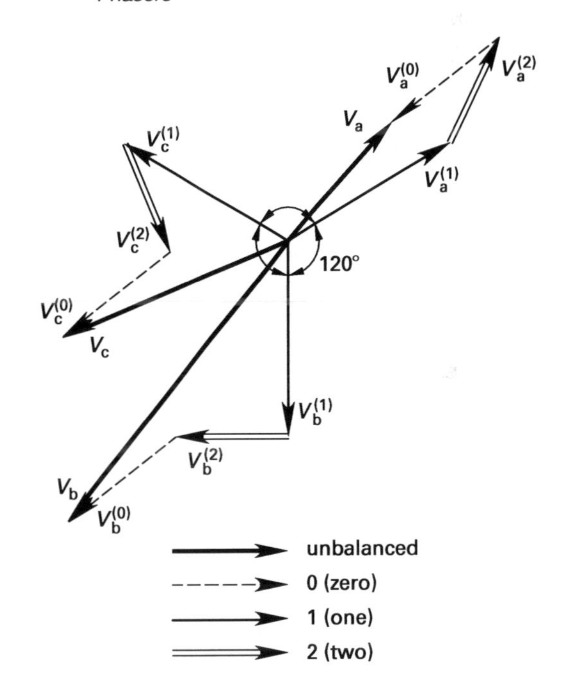

- *rise time*

$$t_r = \frac{\pi - \arccos\zeta}{\omega_d} \quad 41.5$$

- *peak time*

$$t_p = \frac{\pi}{\omega_d} \quad 41.6$$

- *peak gain (fraction overshoot)*

$$M_p = e^{(-\pi\zeta)/\sqrt{1-\zeta^2}} \quad 41.7$$

- *settling time*

$$t_s = \frac{3.91}{\zeta\omega_n} \quad [2\% \text{ criterion}] \quad 41.8$$

$$= \frac{3.00}{\zeta\omega_n} \quad [5\% \text{ criterion}] \quad 41.9$$

- *time constant*

$$\tau = \frac{1}{\zeta\omega_n} \quad 41.10$$

Figure 41.1 *Natural Response*

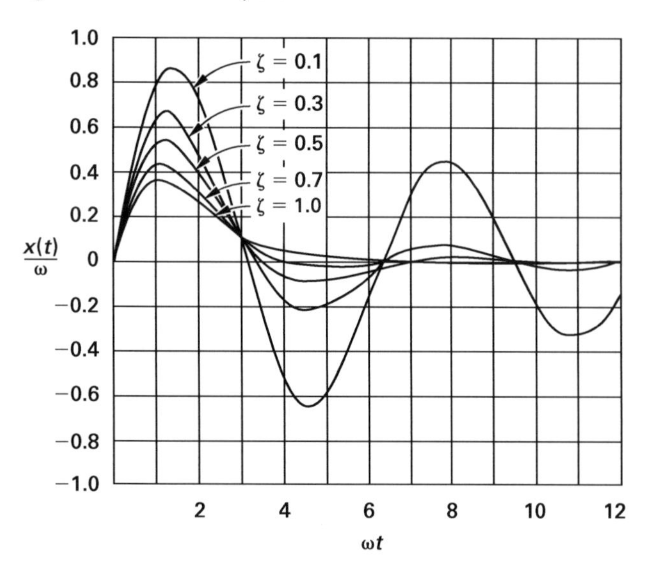

Figure 41.2 *Response to a Unit Step*

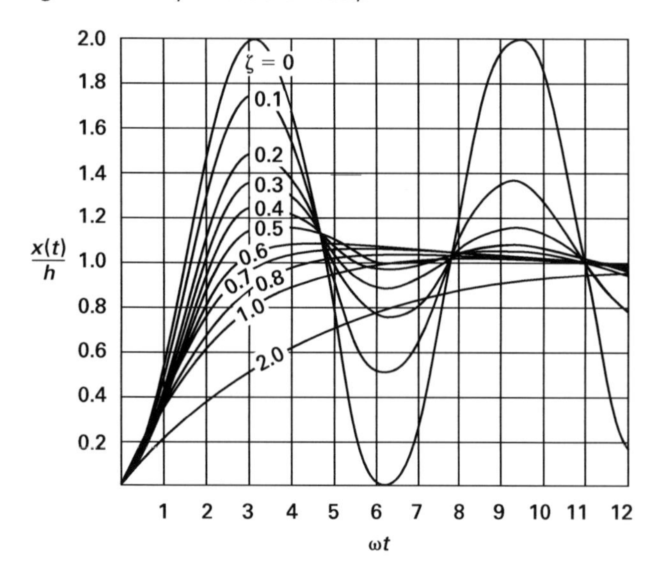

Figure 41.3 *Second-Order Step Time Response Parameters*

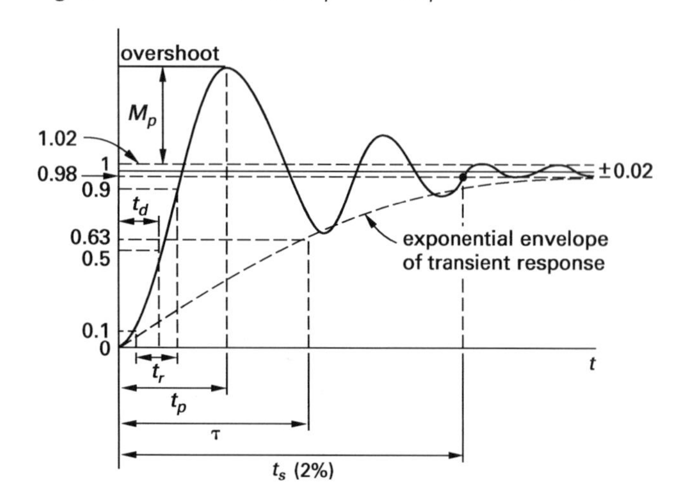

4. FEEDBACK THEORY

- *error transfer function* (*error gain*)

$$E(s) = \mathcal{L}(e(t)) = V_i(s) \pm V_f(s)$$
$$= V_i(s) \pm H(s)V_o(s) \qquad 41.12$$

Figure 41.4 *Feedback System*

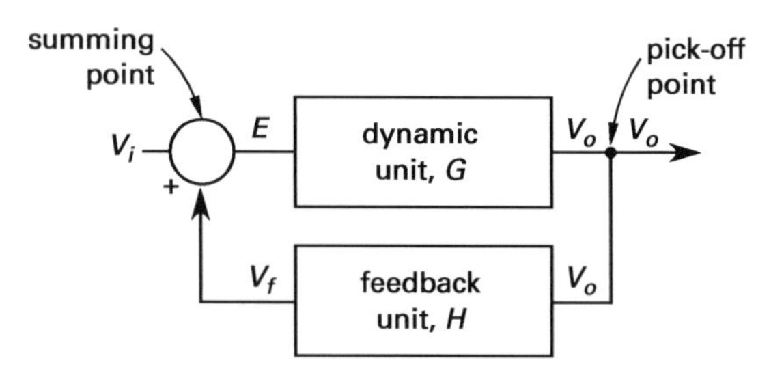

- *forward gain* or *direct gain* ($G(s)$ is normally a complex operator that changes both the magnitude and the phase of the error.)

$$V_o(s) = G(s)E(s) \qquad 41.15$$

The transfer function of the feedback unit is the *reverse transfer function* (*feedback transfer function*, *feedback gain*), $H(s)$, which can be a simple magnitude-changing scalar or a phase-shifting function.

$$V_f(s) = H(s)V_o(s) \qquad 41.16$$

- *feedback ratio* (*primary feedback ratio*)

$$\frac{V_f(s)}{V_i(s)} = \frac{G(s)H(s)}{1 + G(s)H(s)} \quad \text{[negative feedback]} \qquad 41.17$$

$$= \frac{G(s)H(s)}{1 - G(s)H(s)} \quad \text{[positive feedback]} \qquad 41.18$$

The *loop transfer function* (*loop gain* or *open-loop transfer function*) is the gain after going around the loop one time, $\pm G(s)H(s)$.

The *overall transfer function* (*closed-loop transfer function*, *control ratio*, *system function*, *closed-loop gain*), $G_{\text{loop}}(s)$, is the overall transfer function of the feedback system. The quantity $1 + G(s)H(s) = 0$ is the *characteristic equation*.

$$G_{\text{loop}}(s) = \frac{V_o(s)}{V_i(s)}$$

$$= \frac{G(s)}{1 + G(s)H(s)} \quad \text{[negative feedback]} \qquad 41.19$$

$$= \frac{G(s)}{1 - G(s)H(s)} \quad \text{[positive feedback]} \qquad 41.20$$

5. SENSITIVITY

- *sensitivity of the loop transfer function* (with respect to the forward transfer function)

$$S_{G(s)}^{G_{\text{loop}}(s)} = \left(\frac{\Delta G_{\text{loop}}(s)}{\Delta G(s)}\right)\left(\frac{G(s)}{G_{\text{loop}}(s)}\right)$$

$$= \frac{1}{1+G(s)H(s)} \quad \begin{bmatrix}\text{negative}\\ \text{feedback}\end{bmatrix} \qquad 41.22$$

$$= \frac{1}{1-G(s)H(s)} \quad \begin{bmatrix}\text{positive}\\ \text{feedback}\end{bmatrix} \qquad 41.23$$

9. INITIAL AND FINAL VALUES

$$\lim_{t\to 0+} p(t) = \lim_{s\to\infty}(sP(s)) \quad \text{[initial value]} \qquad 41.26$$

$$\lim_{t\to\infty} p(t) = \lim_{s\to 0}(sP(s)) \quad \text{[final value]} \qquad 41.27$$

11. POLES AND ZEROS

A *pole* is a value of s that makes a function, $P(s)$, infinite. Specifically, a pole makes the denominator of $P(s)$ zero. A zero of the function makes the numerator of $P(s)$ (and hence $P(s)$ itself) zero.

12. PREDICTING SYSTEM TIME RESPONSE FROM RESPONSE POLE-ZERO DIAGRAMS

A response pole-zero diagram based on $R(s)$ can be used to predict how the system responds to a specific input.

$$|R| = \frac{K\prod_z |L_z|}{\prod_p |L_p|} = \frac{K\prod_z \text{length}}{\prod_p \text{length}} \qquad 41.32$$

$$\angle R = \sum_p \alpha - \sum_z \beta \qquad 41.33$$

Figure 41.6 *Types of Response Determined by Pole Location*

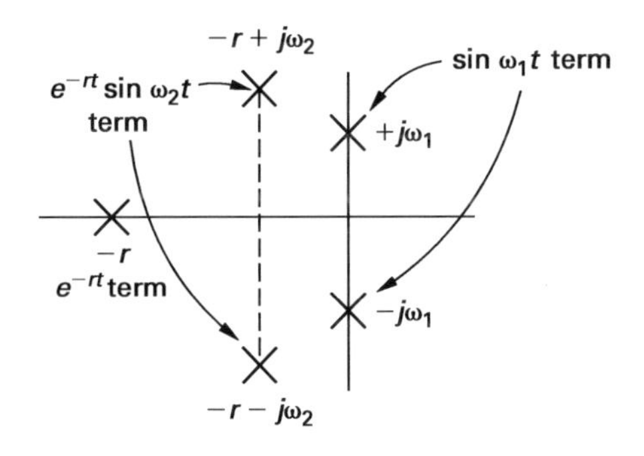

14. GAIN CHARACTERISTIC

- *quality factor*, Q

$$Q = \frac{\omega_n}{\text{BW}} \qquad 41.37$$

Figure 41.8 *Bandwidth*

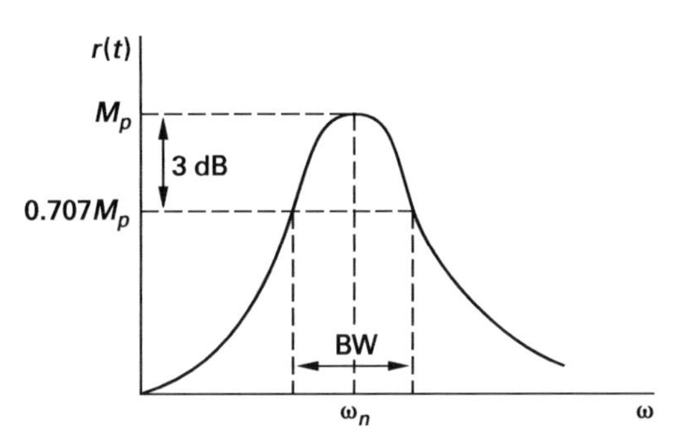

16. STABILITY

The value of the denominator of $T(s)$ is the primary factor affecting stability. When the denominator approaches zero, the system increases without bound. In the typical feedback loop, the denominator is $1 \pm GH$, which can be zero only if $|GH| = 1$. Since $\log 1 = 0$, the requirement for stability is that $\log GH$ must not equal 0 dB.

A negative feedback system will also become unstable if it changes to a positive feedback system, which can occur when the feedback signal is changed in phase more than 180°. Therefore, another requirement for stability is that the phase angle change must not exceed 180°.

17. BODE PLOTS

Bode plots are gain and phase characteristics for the open-loop $G(s)H(s)$ transfer function that are used to determine the *relative stability* of a system.

The *gain margin* is the number of decibels that the open-loop transfer function, $G(s)H(s)$, is below 0 dB.

The *phase margin* is the number of degrees the phase angle is above −180° at the *gain crossover point*.

(See Fig. 41.9.)

18. ROOT-LOCUS DIAGRAMS

A *root-locus diagram* is a pole-zero diagram showing how the poles of $G(s)H(s)$ move when one of the system parameters (e.g., the gain factor) in the transfer function is varied. The diagram gets its name from the need to find the roots of the denominator (i.e., the poles). The locus of points defined by the various poles is a line or curve that can be used to predict *points of instability* or other critical operating points. A point of instability is reached when the line crosses the imaginary axis into the right-hand side of the pole-zero diagram.

Figure 41.9 Gain and Phase Margin Bode Plots

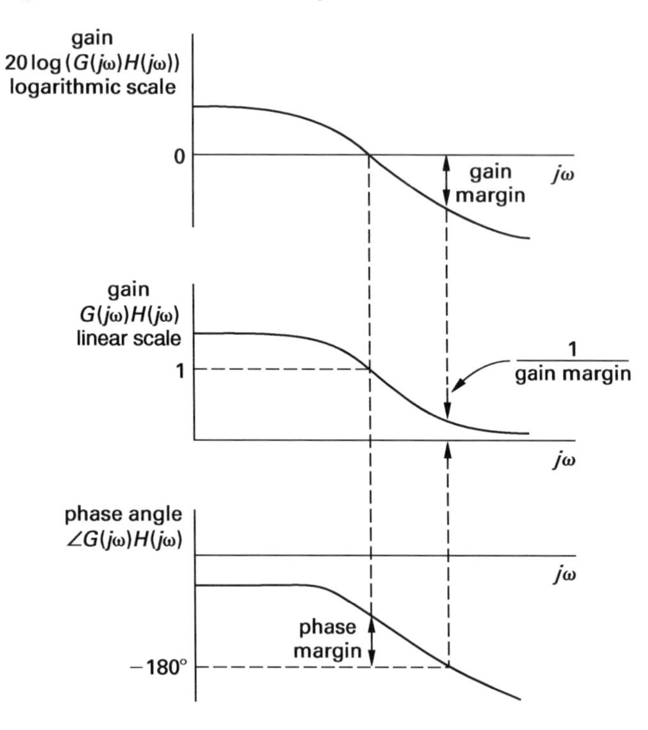

19. HURWITZ TEST

A stable system has poles only in the left half of the s-plane.

- *characteristic equation* (expanded into a polynomial of the form)

$$a_0s^n + a_1s^{n-1} + \cdots + a_{n-1}s + a_n = 0 \quad \textit{41.39}$$

The *Hurwitz stability criterion* requires that all coefficients be present and be of the same sign (which is equivalent to requiring all coefficients to be positive).

20. ROUTH CRITERION

The *Routh criterion*, like the Hurwitz test, uses the coefficients of the polynomial characteristic equation. The Routh-Hurwitz criterion states that the number of sign changes in the first column of the table equals the number of positive (unstable) roots. Therefore, the system will be stable if all entries in the first column have the same sign.

The table is organized in the following manner.

a_0	a_2	a_4	a_6	$\cdots$
a_1	a_3	a_5	a_7	$\cdots$
b_1	b_2	b_3	b_4	$\cdots$
c_1	c_2	c_3	c_4	$\cdots$
$\vdots$	$\vdots$	$\vdots$	$\vdots$	

The remaining coefficients are calculated in the following pattern until all values are zero.

$$b_1 = \frac{a_1a_2 - a_0a_3}{a_1} \quad \textit{41.40}$$

$$b_2 = \frac{a_1a_4 - a_0a_5}{a_1} \quad \textit{41.41}$$

$$b_3 = \frac{a_1a_6 - a_0a_7}{a_1} \quad \textit{41.42}$$

$$c_1 = \frac{b_1a_3 - a_1b_2}{b_1} \quad \textit{41.43}$$

22. APPLICATION TO CONTROL SYSTEMS

Figure 41.10 Typical Feedback Control System

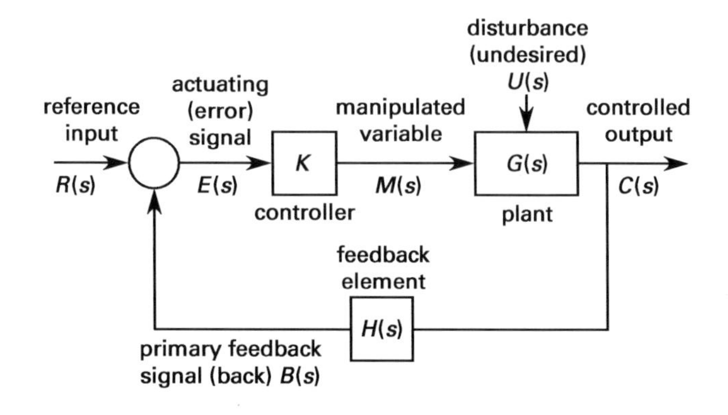

23. CONTROL SYSTEM TYPES/MODES

Figure 41.11 Control System Types

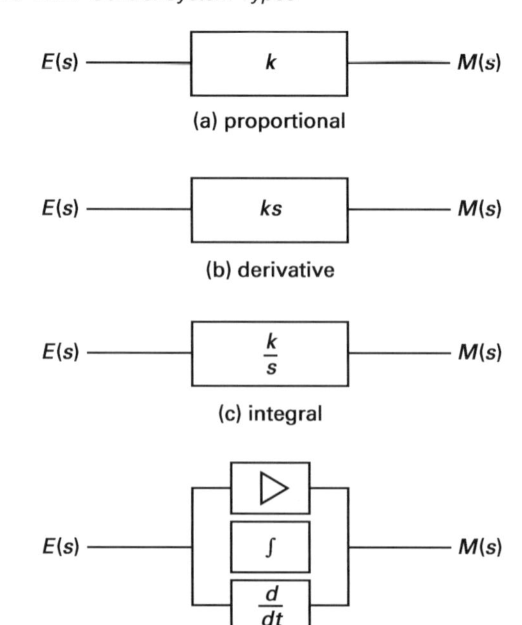

Protection and Safety

EPRM Chapter 42
Protection and Safety

2. POWER SYSTEM GROUNDING

Figure 42.3 *Phasor Diagram of Grounded Phase*

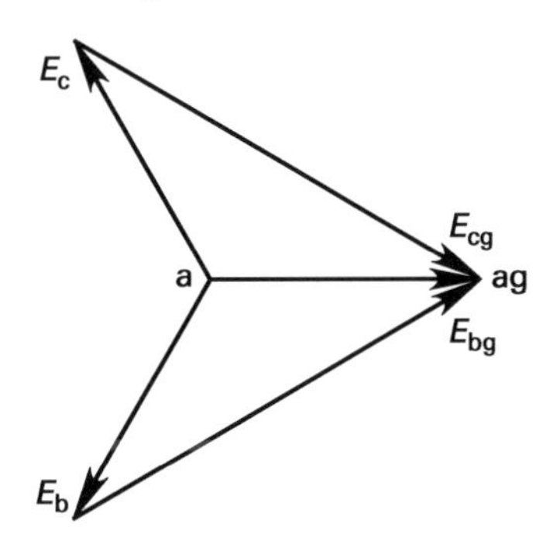

4. RELAYS

(See Table 42.1.)

10. RELAY TYPES

Figure 42.14 *Level Detector Characteristics*

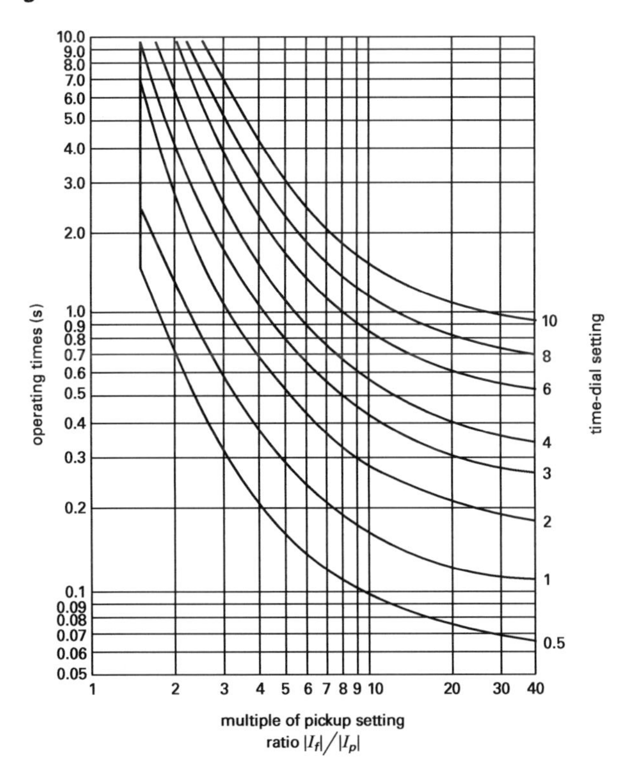

11. UNIVERSAL RELAY EQUATION

(See Fig. 42.17.)

Figure 42.16 *Universal Relay Equation Characteristics*

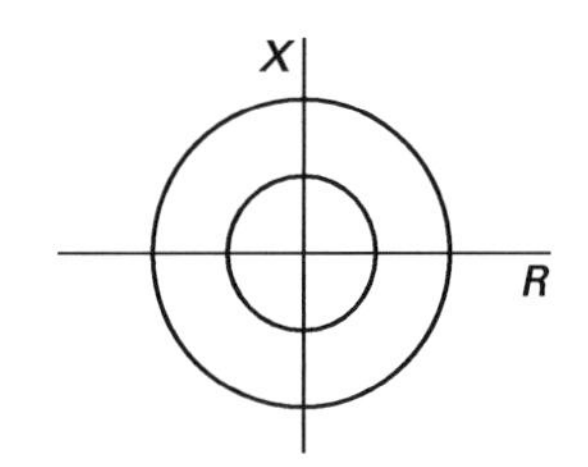

(a) impedance relay

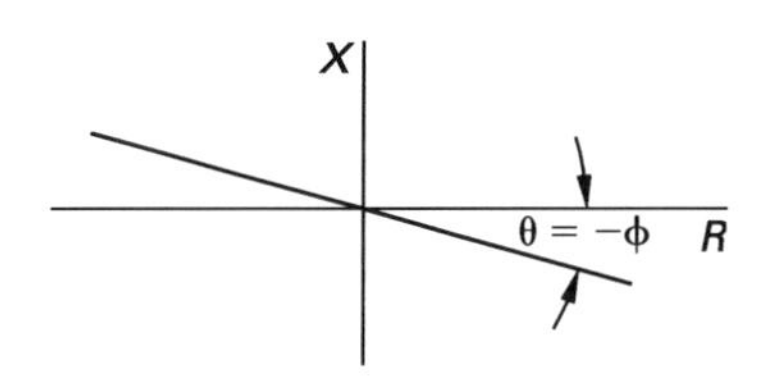

(b) directional relay

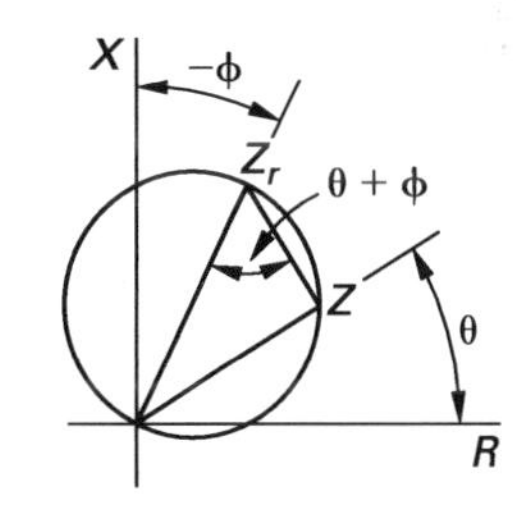

(c) siemens relay

Table 42.1 *ANSI/IEEE Power System Device Function Numbers*

number	title	function
1	master element	places equipment in or out of operation
2	time delay or closing relay	provides desired amount of time delay
3	interlocking relay	allows or stops a given operating sequence
6	starting circuit breaker	connects device to source of voltage
11	multifunction device	combination of functions in one device
12	overspeed device	operates on machine overspeed
14	underspeed device	operates on machine underspeed
21	distance relay	functions when admittance, impedance, or reactance increases or decreases beyond a certain value[a]
24	volts per hertz relay	operates on a given V/Hz ratio[b]
27	undervoltage relay	operates on machine undervoltage
32	directional power relay	operates on power flow in a given direction[c]
46	phase-balance relay	operates when negative phase sequence exceeds a given value[d]
52	AC circuit breaker	operates a circuit breaker
58	rectification relay	operates when a rectifier fails to conduct or block correctly
59	overvoltage relay	operates on overvoltage condition
64	ground detector relay	operates upon failure of insulation to ground
72	DC circuit breaker	operates a circuit breaker
87	differential protection relay	operates upon a quantitative difference between two or more currents or other electrical quantities

[a]A *distance relay* is useful in determining if a fault exists in a given distribution system.
[b]In variable speed devices, the volts per hertz (V/Hz) ratio is adjusted to maintain the current to a machine constant. A constant torque results in a constant speed.
[c]Reverse power on a generator can damage the machine. This relay detects the angular relationship between the current and voltage, which determines the direction of power flow.
[d]This relays checks for negative phase-sequence currents above a given value. Negative phase-sequence currents indicate unbalance phases meaning that a fault exists in the system.

Figure 42.17 *R-X Diagram*

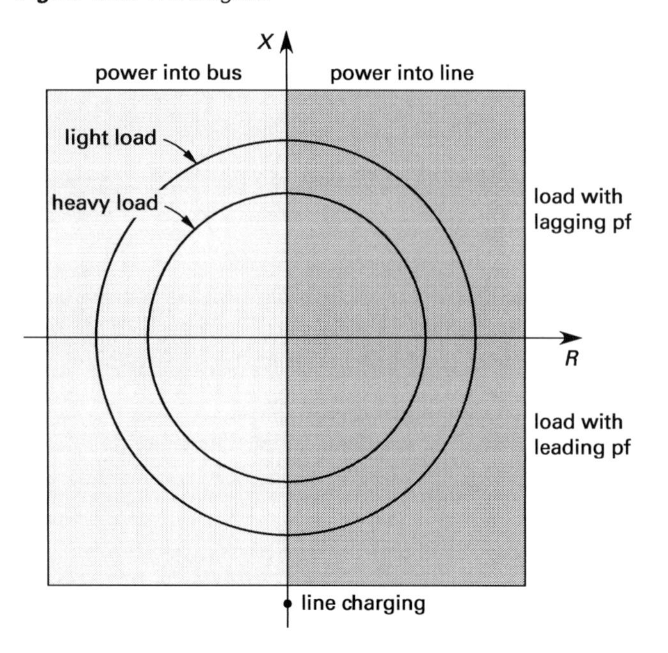

16. ARC-FLASH AND SHOCK HAZARD LEVELS

Figure 42.21 *Approach Boundaries*

arc-flash boundary
limited approach boundary
energized conductor or exposed part
restricted approach boundary

(not to scale)

Table 42.2 *Approach Boundary Limits to Energized Components*

	limited approach boundary		
nominal system voltage range, phase to phase	exposed movable conductor	exposed fixed circuit part	restricted approach boundary
AC < 50 V	not specified	not specified	not specified
50 V ≤ AC ≤ 150 V	3.0 m (10 ft 0 in)	1.0 m (3 ft 6 in)	avoid contact
DC < 100 V	not specified	not specified	not specified
100 V ≤ DC ≤ 300 V	3.0 m (10 ft 0 in)	1.0 m (3 ft 6 in)	avoid contact

Machinery and Devices

EPRM Chapter 43 Rotating DC Machinery

12. TORQUE AND POWER

$$T_{\text{ft-lbf}} = \frac{5252 P_{\text{horsepower}}}{n_{\text{rev/min}}} \quad 43.15$$

$$T_{\text{N·m}} = \frac{1000 P_{\text{kW}}}{\omega_{\text{mech}}} = \frac{9549 P_{\text{kW}}}{n_{\text{rev/min}}} \quad 43.16$$

Equation 43.17 is the general torque expression for a rotating machine with N coils of cross-sectional area A, each carrying current I through a magnetic field of strength B.

$$T = NBAI \cos \omega t \quad 43.17$$

13. SERVICE FACTOR

$$\text{service factor} = \frac{\text{safe load}}{\text{nameplate load}} \quad 43.18$$

16. REGULATION

$$\text{VR} = \frac{V_{\text{nl}} - V_{\text{fl}}}{V_{\text{fl}}} \times 100\% \quad 43.21$$

$$\text{SR} = \frac{n_{\text{nl}} - n_{\text{fl}}}{n_{\text{fl}}} \times 100\% \quad 43.22$$

19. SERIES-WIRED DC MACHINES

$$V = E + I_a(R_a + R_f) \quad 43.26$$

The speed, torque, and current are related by

$$\frac{T_1}{T_2} = \left(\frac{I_{a,1}}{I_{a,2}}\right)^2 \approx \frac{n_1}{n_2} \quad 43.27$$

The torque is

$$T = k'_T \Phi I_a = k_T I_a^2$$
$$= k_T \left(\frac{V}{k_E n + R_a + R_f}\right)^2 \quad 43.28$$

Figure 43.18 *Series-Wired DC Motor Equivalent Circuit*

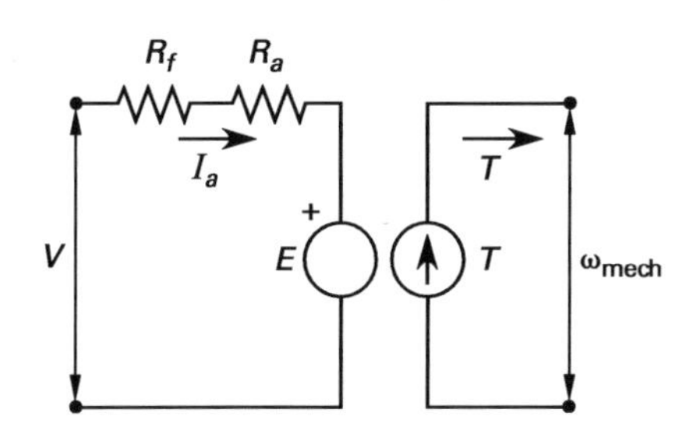

20. SHUNT-WIRED DC MACHINES

$$V = E + I_a R_a = I_f R_f \quad 43.31$$

$$I = I_a + I_f \quad \text{[motor]} \quad 43.32$$

$$I = I_a - I_f \quad \text{[generator]} \quad 43.33$$

$$n = n_{\text{nl}} - k_n T = \frac{V - I_a R_a}{k_E \Phi} \quad 43.34$$

$$\frac{T_1}{T_2} = \frac{I_{a,1}}{I_{a,2}} \quad 43.36$$

Figure 43.19 *Equivalent Circuits*

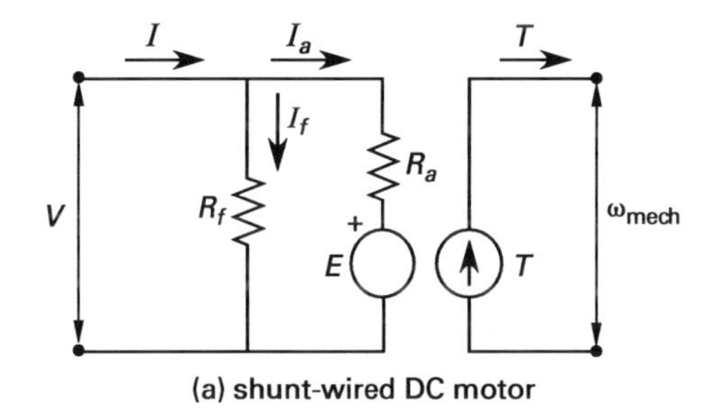

(a) shunt-wired DC motor

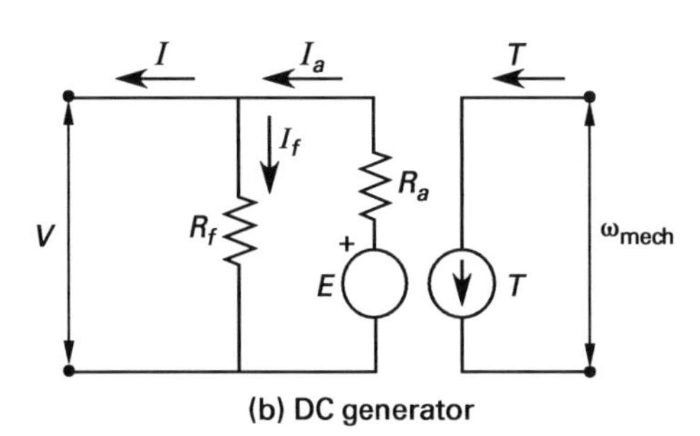

(b) DC generator

22. VOLTAGE-CURRENT CHARACTERISTICS FOR DC GENERATORS

Figure 43.20 DC Generator Voltage-Current Characteristics

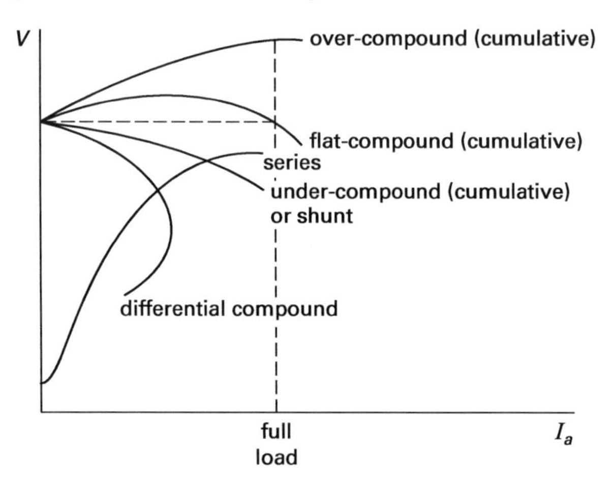

23. TORQUE CHARACTERISTICS FOR DC MOTORS

Figure 43.21 DC Motor Torque Characteristics

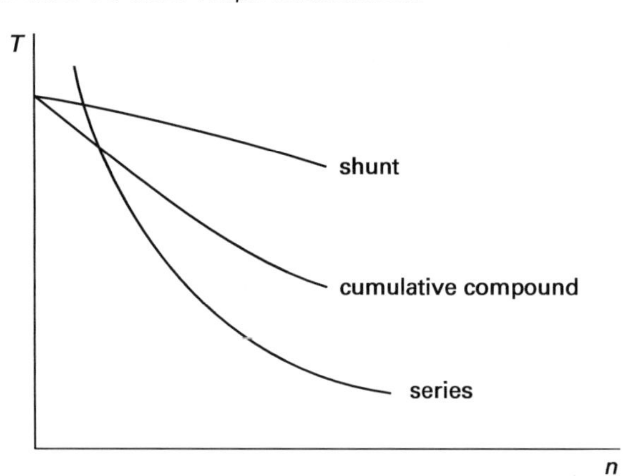

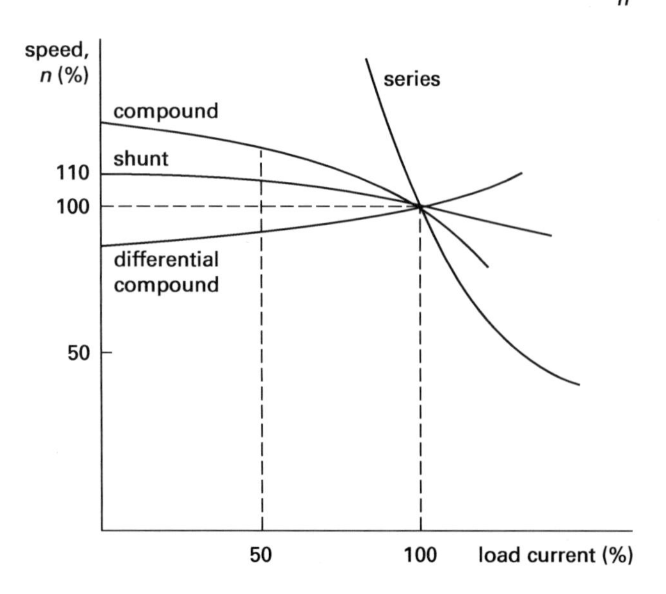

EPRM Chapter 44 Rotating AC Machinery

1. ROTATING MACHINES

$$V_p = \begin{cases} V & \text{[delta-wired]} \\ \dfrac{V}{\sqrt{3}} & \text{[wye-wired]} \end{cases} \quad 44.1$$

$$T_p = \frac{T_t}{3} \quad 44.2$$

$$P_p = \frac{P_t}{3} \quad 44.3$$

$$S_p = \frac{S_t}{3} \quad 44.4$$

2. TORQUE AND POWER

$$T_{\text{ft-lbf}} = \frac{5252 P_{\text{horsepower}}}{n_{\text{rpm}}} \quad 44.5$$

$$T_{\text{N·m}} = \frac{1000 P_{\text{kW}}}{\omega_{\text{mech}}} = \frac{9549 P_{\text{kW}}}{n_{\text{rpm}}} \quad 44.6$$

Equation 44.7 is the general torque expression for a rotating machine with N coils of cross-sectional area A, each carrying current I through a magnetic field of strength B.

$$T = NBAI \cos \omega t \quad 44.7$$

3. SERVICE FACTOR

$$\text{service factor} = \frac{\text{safe load}}{\text{nameplate load}} \quad 44.8$$

6. REGULATION

$$\text{VR} = \frac{V_{\text{nl}} - V_{\text{fl}}}{V_{\text{fl}}} \times 100\% \quad 44.11$$

$$\text{SR} = \frac{n_{\text{nl}} - n_{\text{fl}}}{n_{\text{fl}}} \times 100\% \quad 44.12$$

7. NO-LOAD CONDITIONS

The meaning of the term *no load* is different for generators and motors.

$$I = 0; \quad I_f \neq 0; \quad I_a = I_f \quad \text{[generator]} \quad 44.13$$

$$I \neq 0; \quad I_f = I; \quad I_a = 0 \quad \text{[motor]} \quad 44.14$$

8. PRODUCTION OF AC POTENTIAL

$$V = \frac{V_{\text{max}}}{\sqrt{2}} = \frac{\omega NAB}{\sqrt{2}} = \frac{p\omega_{\text{mech}} NAB}{2\sqrt{2}} = \frac{\pi npNAB}{60\sqrt{2}} \quad \text{[effective]} \qquad 44.16$$

$$n_s = \frac{120f}{p} = \frac{60\omega_{\text{mech}}}{2\pi} = \frac{60\omega}{\pi p} \quad \left[\begin{array}{c}\text{synchronous}\\ \text{speed}\end{array}\right] \qquad 44.17$$

$$f = \frac{1}{T} = \frac{\omega}{2\pi} = \frac{pn_s}{120} \qquad 44.18$$

13. SYNCHRONOUS MACHINE EQUIVALENT CIRCUIT

$$\mathbf{E} = \mathbf{V}_p + (R_a + jX_s)\mathbf{I}_a \approx \mathbf{V}_p + jX_s\mathbf{I}_a \quad \text{[alternator]} \qquad 44.22$$

$$\mathbf{V}_p = \mathbf{E} + (R_a + jX_s)\mathbf{I}_a \approx \mathbf{E} + jX_s\mathbf{I}_s \quad \text{[motor]} \qquad 44.23$$

$$P_p = T_p\omega_{\text{mech}} = VI\cos\theta = \left(\frac{VE}{X_s}\right)\sin\delta \qquad 44.24$$

Figure 44.5 Synchronous Motor Equivalent Circuit

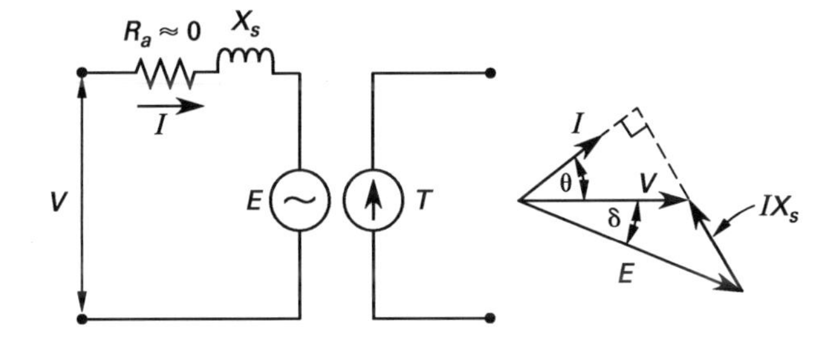

15. INDUCTION MOTORS

Slip (in rpm) is the difference between actual and synchronous speeds.

$$s = \frac{n_s - n}{n_s} = \frac{\omega_{\text{mech},s} - \omega_{\text{mech}}}{\omega_{\text{mech},s}} \qquad 44.27$$

16. INDUCTION MOTOR EQUIVALENT CIRCUIT

Using an adjusted voltage, V_{adj}, simplifies the model, as shown in Fig. 44.8(b). The relationship between the applied terminal voltage, V_1, and the adjusted voltage is

$$\mathbf{V}_{\text{adj}} = \mathbf{V}_1 - \mathbf{I}_{\text{nl}}(R_1 + jX_1) \qquad 44.28$$

$$V_{\text{adj}} \approx V_1 - I_{\text{nl}}\sqrt{R_1^2 + X_1^2} \qquad 44.29$$

Figure 44.8 Equivalent Circuits of an Induction Motor

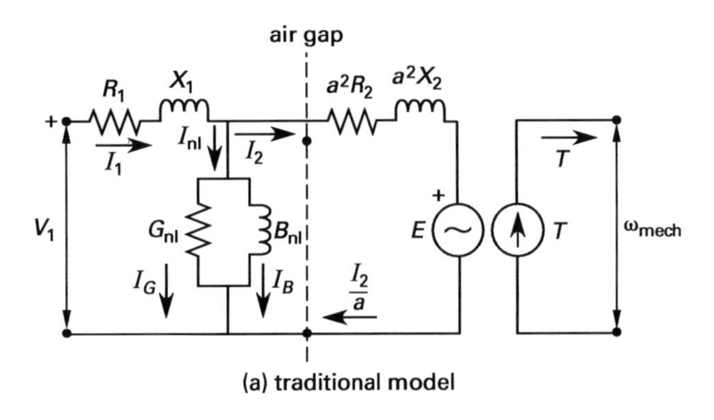

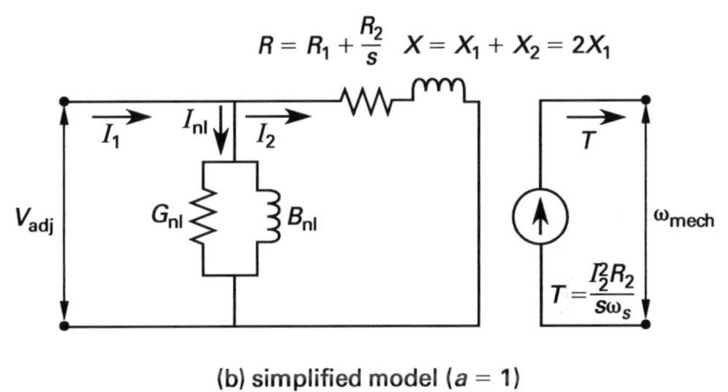

17. OPERATING CHARACTERISTICS OF INDUCTION MOTORS

Figure 44.9 Characteristic Curves for an Induction Motor

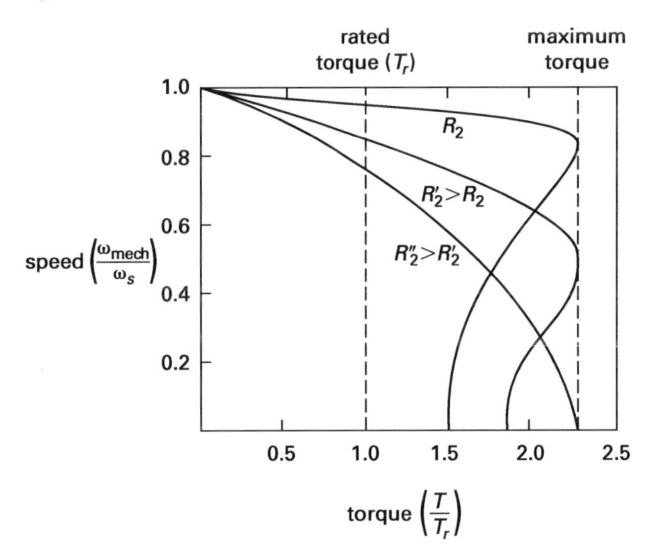

21. POWER TRANSFER IN INDUCTION MOTORS

$$\text{input power} = V_1 I_1 \cos\theta \qquad 44.48$$

$$\text{stator copper losses} = I_1^2 R_1 \qquad 44.49$$

$$\text{rotor input power} = \frac{I_2^2 R_2}{s} \qquad 44.50$$

$$\text{rotor copper losses} = I_2^2 R_2 \qquad 44.51$$

$$\text{electrical power delivered} = I_2^2 R_2 \left(\frac{1-s}{s}\right) \qquad 44.52$$

$$\text{shaft output power} = T\omega_{\text{mech}} \qquad 44.53$$

Figure 44.10 *Induction Motor Power Transfer*

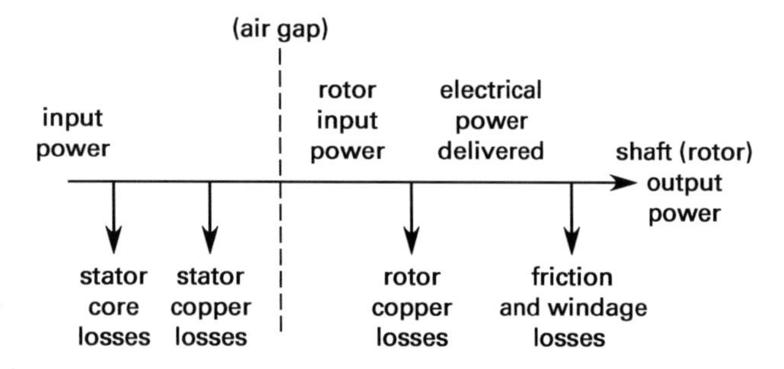

Electronics

EPRM Chapter 45
Electronics Fundamentals

1. OVERVIEW

- *mass action law*

$$n_i^2 = np \qquad 45.1$$

2. SEMICONDUCTOR MATERIALS

- *density of electron-hole pairs in intrinsic materials*

$$n_i^2 = A_0 T^3 e^{-E_{G0}/\kappa T} \qquad 45.2$$

$$n_i^2 = N_c N_v e^{-E_G/\kappa T} \qquad 45.3$$

- *law of electrical neutrality*

$$N_A + n = N_D + p \qquad 45.5$$

- *concentration of electrons in a p-type material*

$$n = \frac{n_i^2}{N_A} \qquad 45.6$$

- *concentration of holes in an n-type material*

$$p = \frac{n_i^2}{N_D} \qquad 45.7$$

5. AMPLIFIERS

(See Fig. 45.2.)

7. LOAD LINE AND QUIESCENT POINT CONCEPT

Determine the load line for a generic transistor amplifier as follows.

step 1: For the configuration provided, label the x-axis on the *output characteristic curves* with the appropriate voltage. (For a BJT, this is V_{CE} or V_{CB}. For a FET, this is V_D.)

step 2: Label the y-axis as the output current. (For a BJT, this is I_C. For a FET, this is I_D.)

step 3: Redraw the circuit with all three terminals of the transistor open. Label the terminals. (For a BJT, these are base, emitter, and collector. For a FET, these are gate, source, and drain.) Label the current directions all pointing inward, toward the amplifier. (For a BJT, these are I_B, I_E, and I_C. For a FET, these are I_D and I_S.)

Figure 45.2 *General Amplifier*

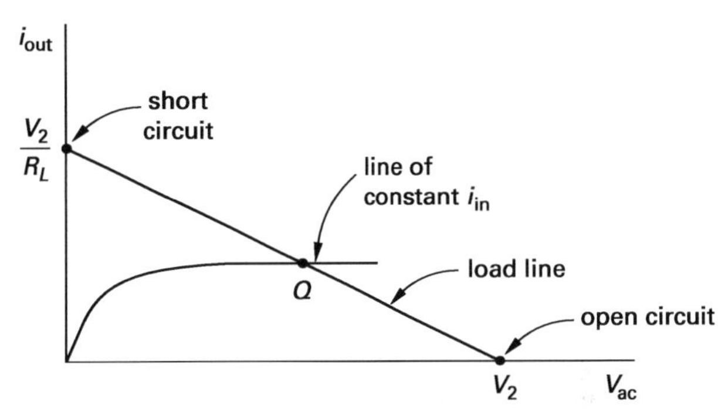

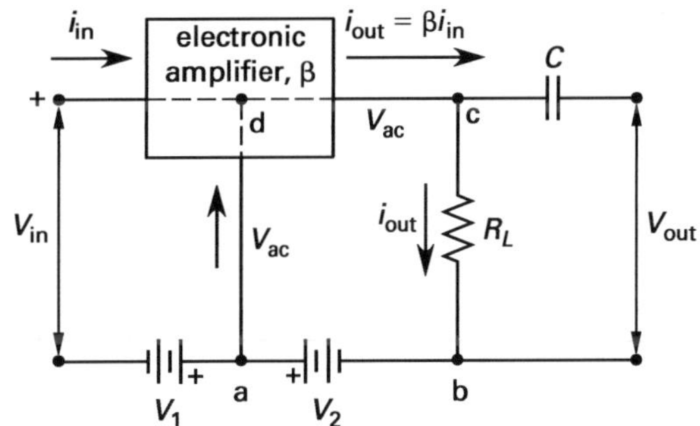

step 4: Perform KVL analysis in the output loop. (For a BJT, this is the collector loop. For a FET, this is the drain loop.) The transistor voltage determined is a point on the x-axis with the output current equal to zero. Plot the point.

step 5: Redraw the circuit with all three terminals of the transistor shorted. Label as in step 3.

step 6: Use Ohm's law, or another appropriate method, in the output loop to determine the current. (For the BJT, this is the collector current. For the FET, this is the drain current.) The transistor current determined is a point on the y-axis with the applicable voltage in step 1 equal to zero. Plot the point.

step 7: Draw a straight line between the two points. This is the DC load line.

8. *pn* JUNCTIONS

An ideal *pn* junction, excluding the breakdown region, is governed by

$$I_{pn} = I_s(e^{qV_{pn}/\kappa T} - 1) \qquad 45.15$$

9. DIODE PERFORMANCE CHARACTERISTICS

Figure 45.9 Semiconductor Diode Characteristics and Symbol

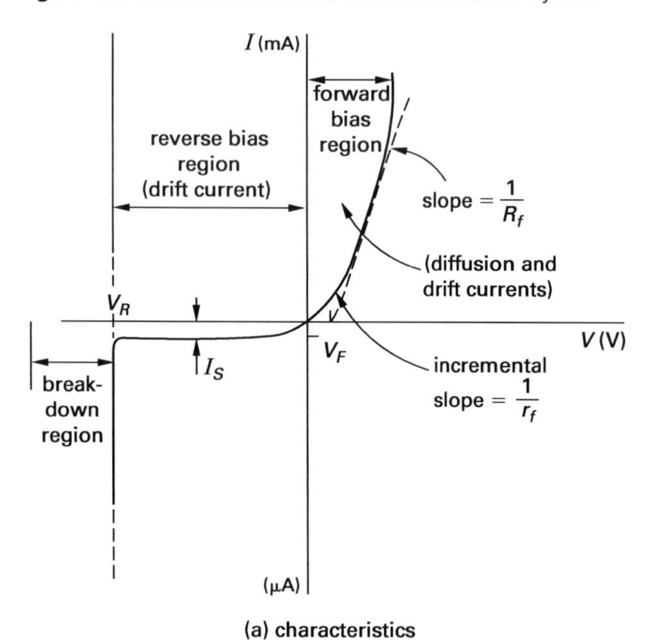

(a) characteristics

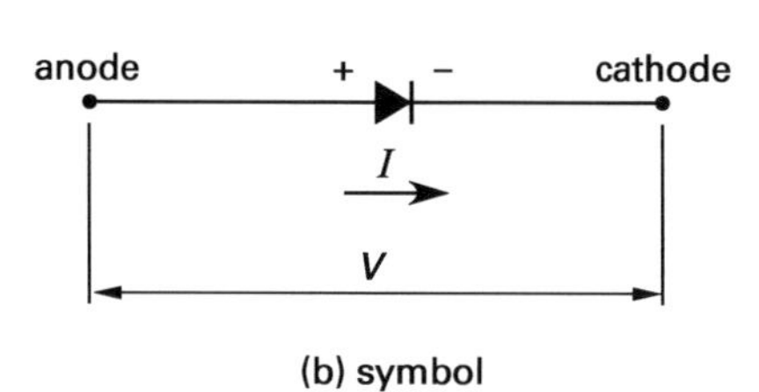

(b) symbol

For practical junctions (*diodes* or *rectifiers*),

$$I = I_s(e^{qV/\eta\kappa T} - 1) = I_s(e^{V/\eta V_T} - 1) \qquad 45.16$$

- *voltage equivalent of temperature*

$$V_T = \frac{\kappa T}{q} = \frac{D_p}{\mu_p} = \frac{D_n}{\mu_n} \qquad 45.17$$

The temperature has the effect of doubling the saturation current every 10°C.

$$\frac{I_{s2}}{I_{s1}} = (2)^{T_2 - T_1/10°\text{C}} \qquad 45.18$$

Figure 45.10 Diode Equivalent Circuit

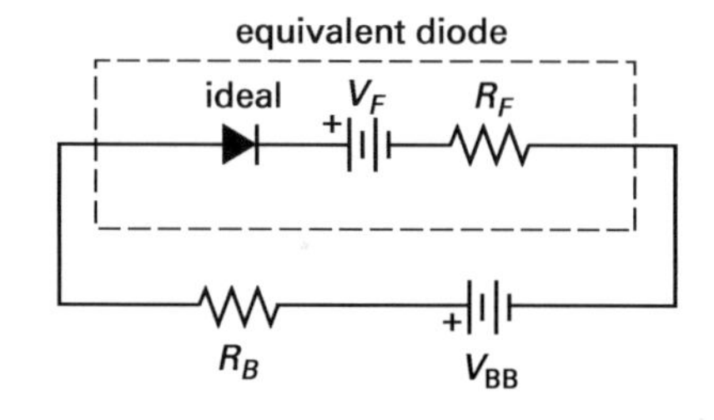

- *dynamic forward resistance, r_f*

$$R_f = r_f = \frac{\eta V_T}{I_D} \qquad 45.19$$

10. DIODE LOAD LINE

Figure 45.11 Diode Load Line

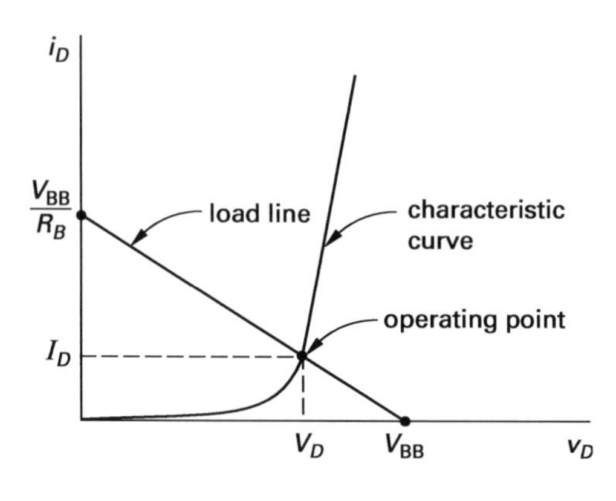

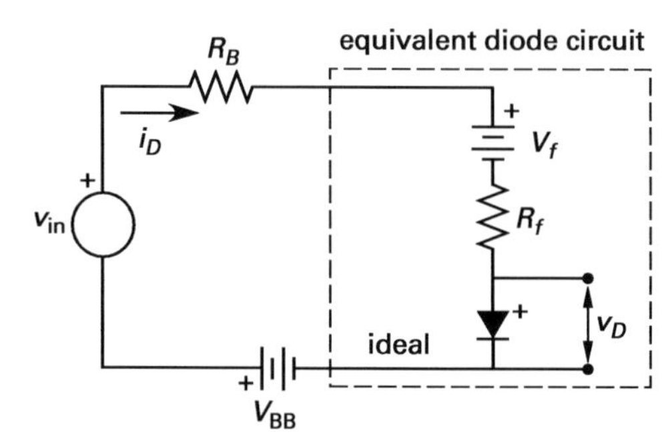

11. DIODE PIECEWISE LINEAR MODEL

Figure 45.12 Piecewise Linear Model

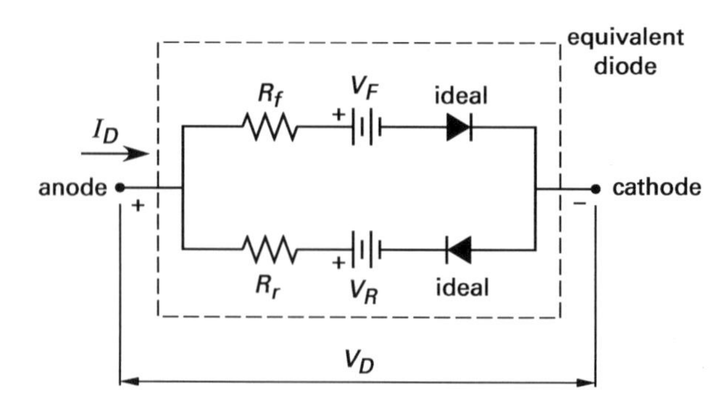

16. PHOTODIODES AND LIGHT-EMITTING DIODES

- *emitted wavelength*

$$\lambda = \frac{hc}{E_G} \qquad 45.26$$

17. SILICON-CONTROLLED RECTIFIERS

Figure 45.17 Silicon-Controlled Rectifier

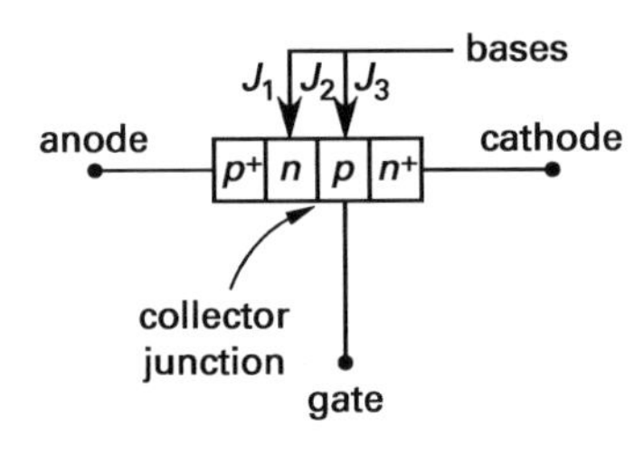

(a) conceptual construction

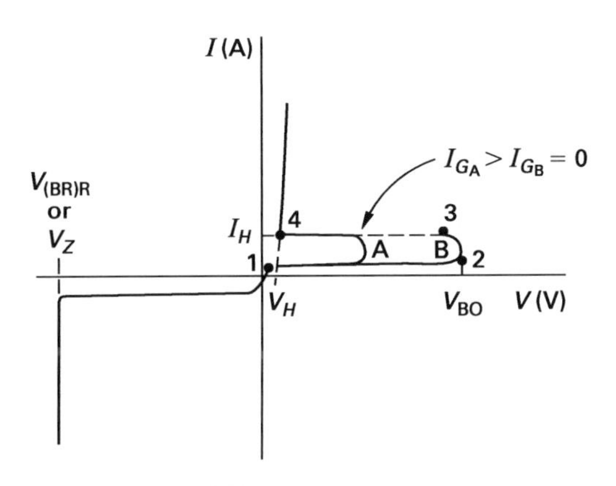

(b) characteristics

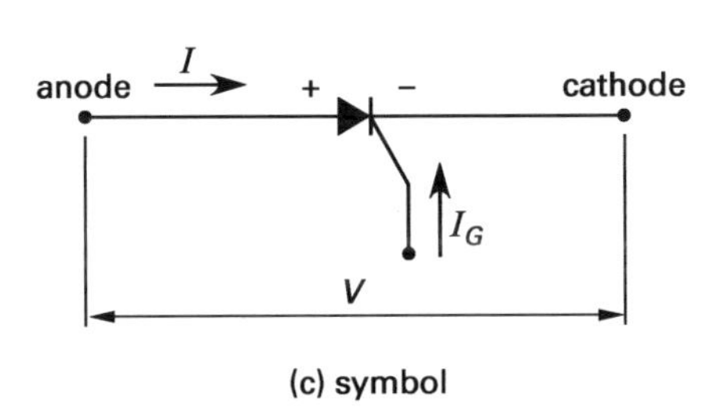

(c) symbol

EPRM Chapter 46 Junction Transistors

1. TRANSISTOR FUNDAMENTALS

(See Fig. 46.1.)

2. BJT TRANSISTOR PERFORMANCE CHARACTERISTICS

(See Fig. 46.3.)

3. BJT TRANSISTOR PARAMETERS

$$I_E = I_C + I_B \qquad 46.2$$

$$\beta_{DC} = \frac{I_C}{I_B} = \frac{\alpha_{DC}}{1 - \alpha_{DC}} \qquad 46.3$$

Figure 46.1 Bipolar Junction Transistor

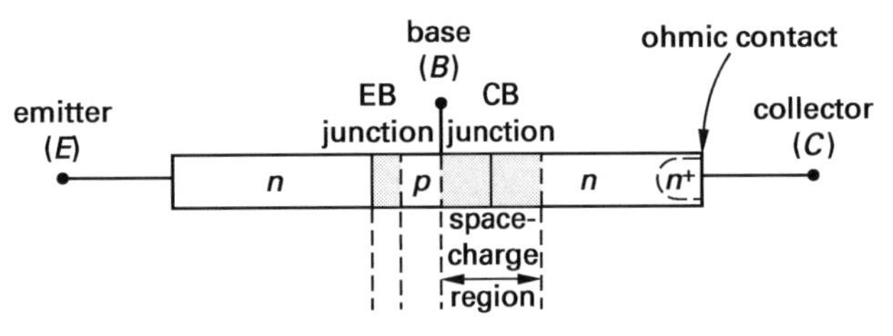

(a) conceptual construction

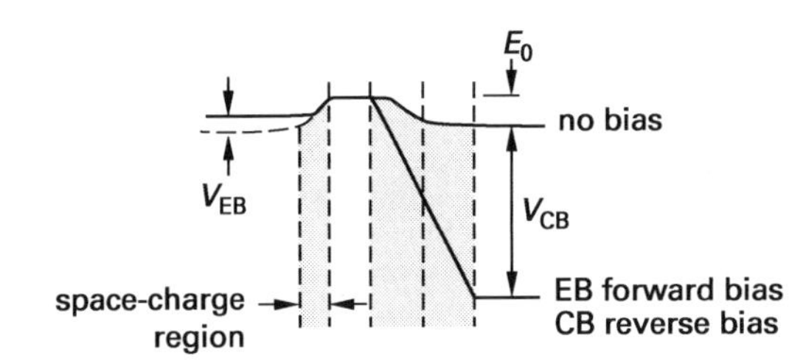

(space-charge regions shown for active region biasing)

(b) electron energy level

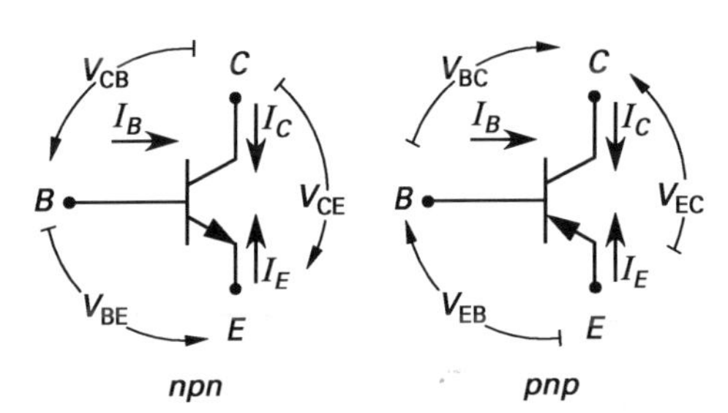

(all currents shown in positive direction)
(all voltages shown for active region biasing)

(c) symbol

$$\alpha_{DC} = \frac{I_C}{I_E} = \frac{\beta_{DC}}{1 + \beta_{DC}} \qquad 46.4$$

The difference between β_{ac} and β_{DC} is very small, and the two are not usually distinguished.

$$\beta_{ac} = \frac{\Delta I_C}{\Delta I_B} = \frac{i_C}{i_B} \qquad 46.5$$

$$\alpha_{ac} = \frac{\Delta I_C}{\Delta I_E} = \frac{i_C}{i_E} \qquad 46.6$$

I_{CBO} is the thermal current at the collector-base junction.

$$I_C = I_E - I_B \qquad 46.7$$

$$I_C = \alpha I_E - I_{CBO} \approx \alpha I_E \qquad 46.8$$

Figure 46.3 *BJT Output Characteristics*

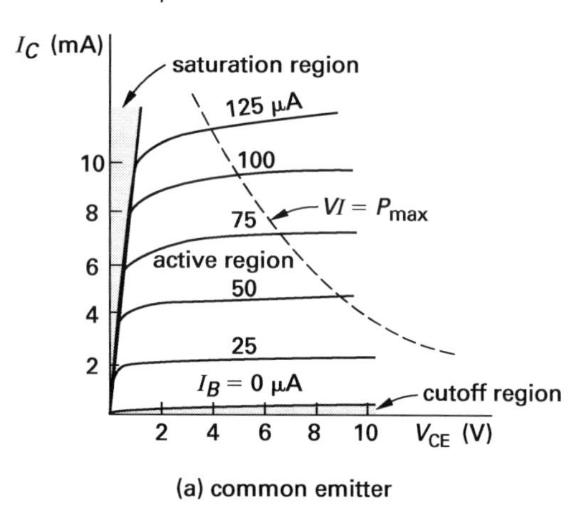

(a) common emitter

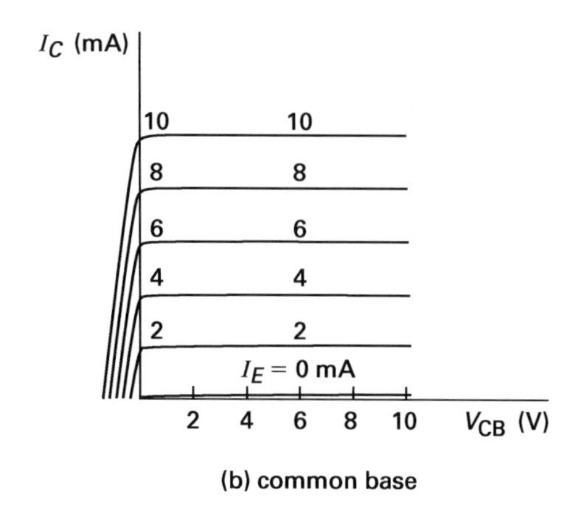

(b) common base

- *equivalence of symbols*

$$\alpha_{DC} = h_{FB} \tag{46.9}$$

$$\alpha_{ac} = h_{fb} \tag{46.10}$$

$$\beta_{DC} = h_{FE} \tag{46.11}$$

$$\beta_{ac} = h_{fe} \tag{46.12}$$

6. BJT LOAD LINE

Figure 46.7 *Common Emitter Load Line and Quiescent Point*

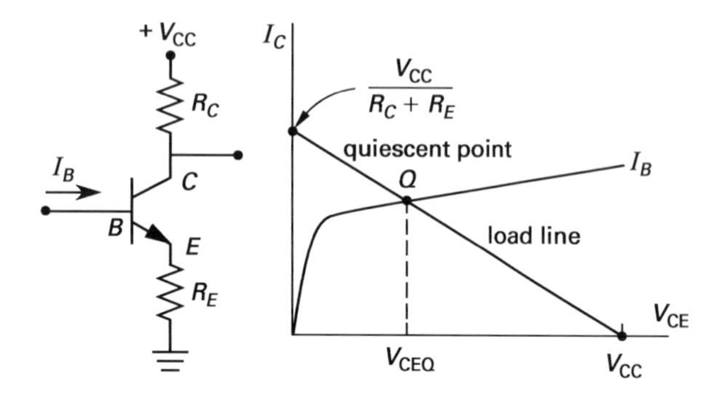

7. AMPLIFIER GAIN AND POWER

$$A_V = \frac{\Delta V_{out}}{\Delta V_{in}} = \frac{v_{out}}{v_{in}} = \beta A_R \tag{46.17}$$

$$A_I = \frac{\Delta I_{out}}{\Delta I_{in}} = \frac{i_{out}}{i_{in}} = \beta \tag{46.18}$$

$$A_R = \frac{Z_{out}}{Z_{in}} = \frac{A_V}{\beta} \tag{46.19}$$

$$A_P = \frac{P_{out}}{P_{in}} = \beta^2 A_R = A_I A_V \tag{46.20}$$

9. EQUIVALENT CIRCUIT REPRESENTATION AND MODELS

(See Fig. 46.8.)

h_i input impedance with output shorted (Ω)
h_r reverse transfer voltage ratio with input open (dimensionless)
h_f forward transfer current ratio with output shorted (dimensionless)
h_o output admittance with input open (S)

Table 46.2 *Equivalent Circuit Parameters*

symbol	common emitter	common collector	common base
h_{11}, h_{ie}	h_{ie}	h_{ic}	$\frac{h_{ib}}{1+h_{fb}}$
h_{12}, h_{re}	h_{re}	$1 - h_{rc}$	$\frac{h_{ib}h_{ob}}{1+h_{fb}} - h_{rb}$
h_{21}, h_{fe}	h_{fe}	$-1 - h_{fc}$	$\frac{-h_{fb}}{1+h_{fb}}$
h_{22}, h_{oe}	h_{oe}	h_{oc}	$\frac{h_{ob}}{1+h_{fb}}$
h_{11}, h_{ib}	$\frac{h_{ie}}{1+h_{fe}}$	$\frac{-h_{ic}}{h_{fc}}$	h_{ib}
h_{12}, h_{rb}	$\frac{h_{ie}h_{oe}}{1+h_{fe}} - h_{re}$	$h_{rc} - \frac{h_{ic}h_{oc}}{h_{fc}} - 1$	h_{rb}
h_{21}, h_{fb}	$\frac{-h_{fe}}{1+h_{fe}}$	$\frac{-1-h_{fc}}{h_{fc}}$	h_{fb}
h_{22}, h_{ob}	$\frac{h_{oe}}{1+h_{fe}}$	$\frac{-h_{oc}}{h_{fc}}$	h_{ob}
h_{11}, h_{ic}	h_{ie}	h_{ic}	$\frac{h_{ib}}{1+h_{fb}}$
h_{12}, h_{rc}	$1 - h_{re}$	h_{rc}	1
h_{21}, h_{fc}	$-1 - h_{fe}$	h_{fc}	$\frac{-1}{1+h_{fb}}$
h_{22}, h_{oc}	h_{oe}	h_{oc}	$\frac{h_{ob}}{1+h_{fb}}$

10. APPROXIMATE TRANSISTOR MODELS

The values of h_r and h_o are very small. The simplified models of Table 46.5 are obtained by ignoring these two parameters.

Figure 46.8 *Transistors as Two-Port Networks*

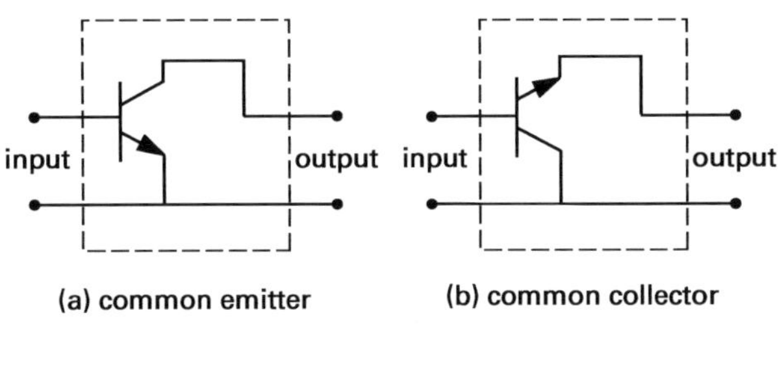

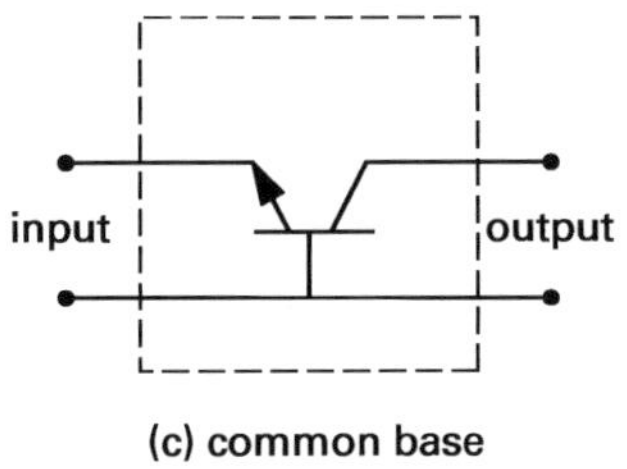

Table 46.5 *BJT Simplified Equivalent Circuits*

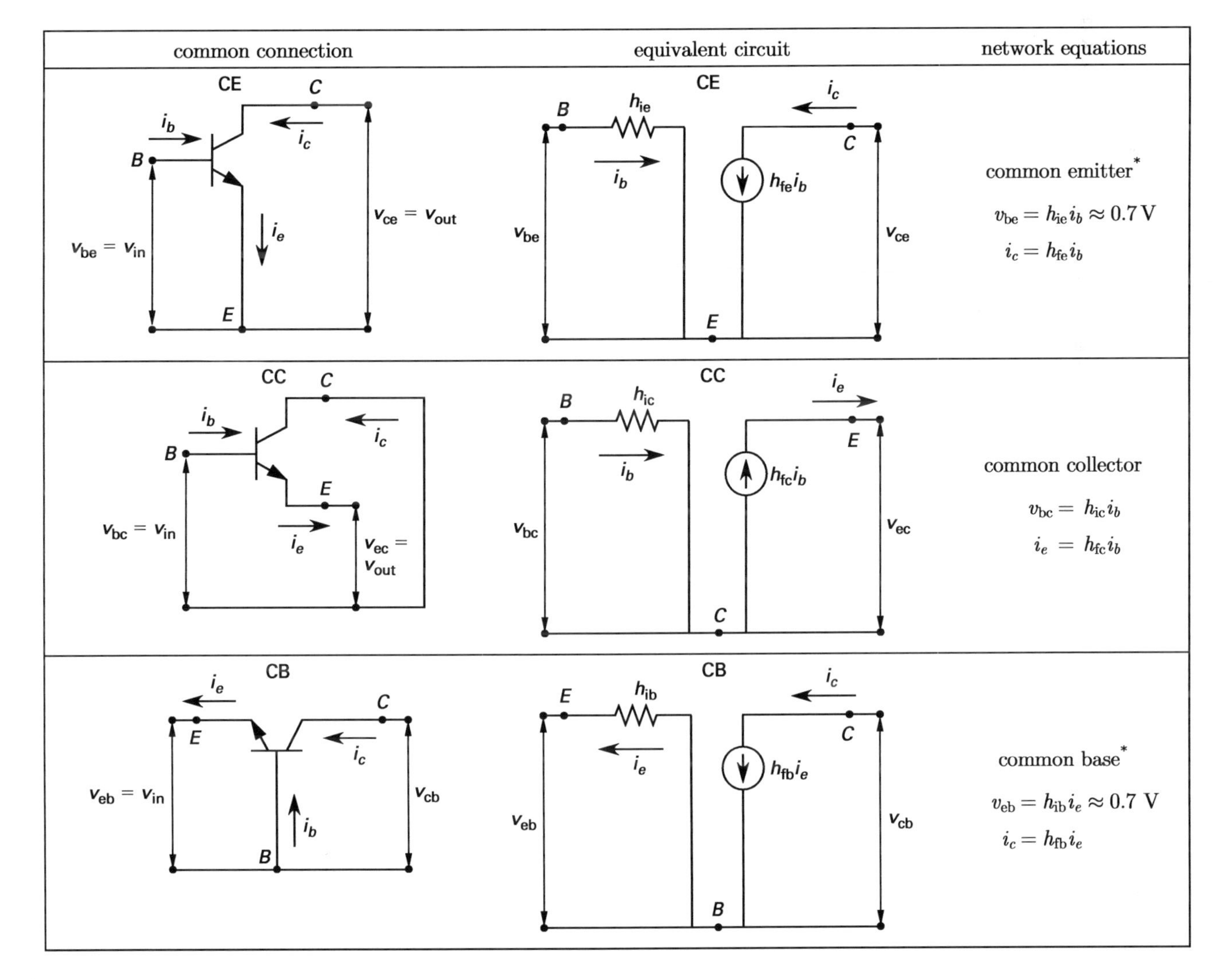

*Germanium transistors are *pnp* types. $|v_{be}| = |v_{eb}| = 0.3$ V for germanium.

EPRM Chapter 47
Field Effect Transistors

2. JFET CHARACTERISTICS

The term "pinchoff voltage" and the symbol V_P are ambiguous, as the actual pinchoff voltage in a circuit depends on the gate-source voltage, V_{GS}. When V_{GS} is zero, the pinchoff voltage is represented unambiguously by V_{P0}. (See Fig. 47.1.)

$$V_P = V_{P0} + V_{GS} \tag{47.1}$$

- *Shockley's equation*

$$I_D = I_{DSS}\left(1 - \frac{V_{GS}}{V_P}\right)^2 \tag{47.2}$$

Figure 47.1 *Junction Field-Effect Transistor*

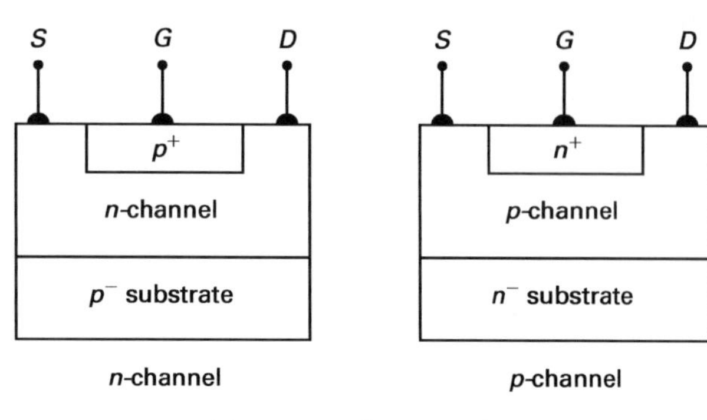

(a) conceptual construction

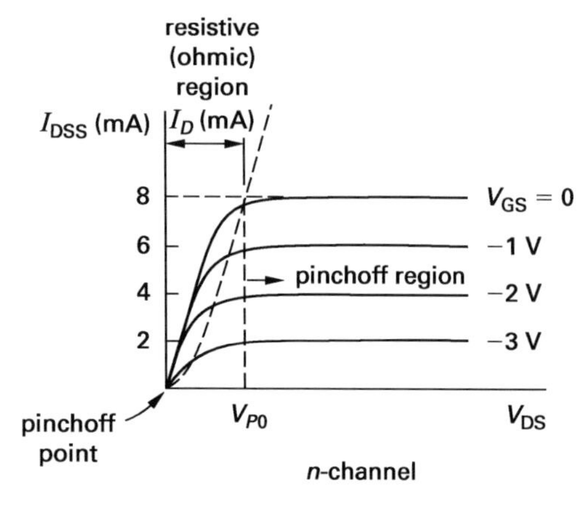

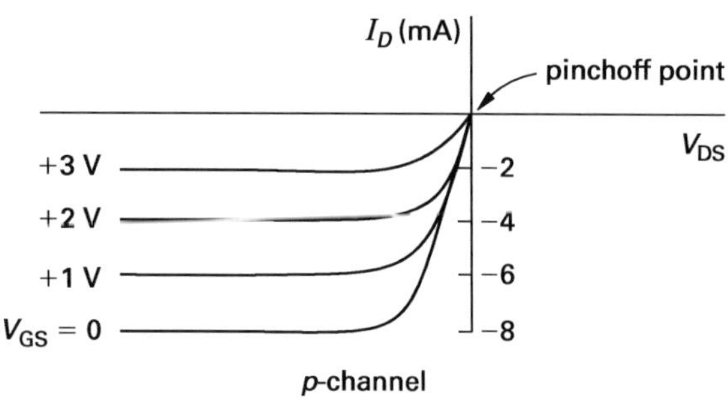

(b) characteristics

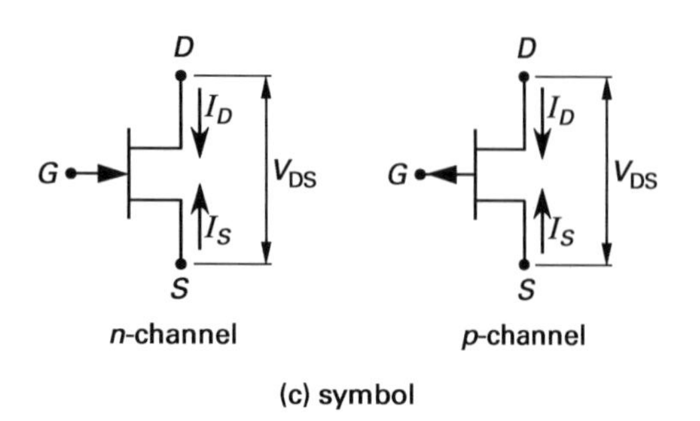

(c) symbol

- *transconductance, g_m*

$$
\begin{aligned}
g_m &= \frac{\Delta I_D}{\Delta V_{GS}} = \frac{i_D}{v_{GS}} \\
&= \left(\frac{-2I_{DSS}}{V_P}\right)\left(1-\frac{V_{GS}}{V_P}\right) \\
&= g_{mo}\left(1-\frac{V_{GS}}{V_P}\right) \\
&\approx \frac{A_V}{R_{out}}
\end{aligned}
\qquad 47.3
$$

The drain-source resistance can be obtained from the slope of the V_{GS} characteristic in Fig. 47.1(b).

$$
r_d = r_{DS} = \frac{\Delta V_{DS}}{\Delta I_D} = \frac{v_{DS}}{i_D} \qquad 47.4
$$

3. JFET BIASING

- *load line equation*

$$
\begin{aligned}
V_{DD} &= I_S(R_D + R_S) + V_{DS} \\
&= I_D(R_D + R_S) + V_{DS}
\end{aligned}
\qquad 47.5
$$

At the quiescent point, $V_{in} = 0$. From Kirchhoff's voltage law around the input loop,

$$
V_{GS} = -I_S R_S = -I_D R_S \qquad 47.6
$$

Figure 47.3 *Self-Biasing JFET Circuit*

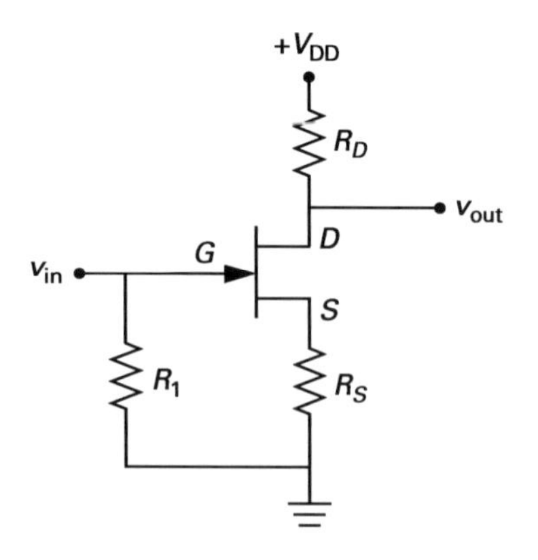

4. FET MODELS

Figure 47.4 *FET Equivalent Circuit*

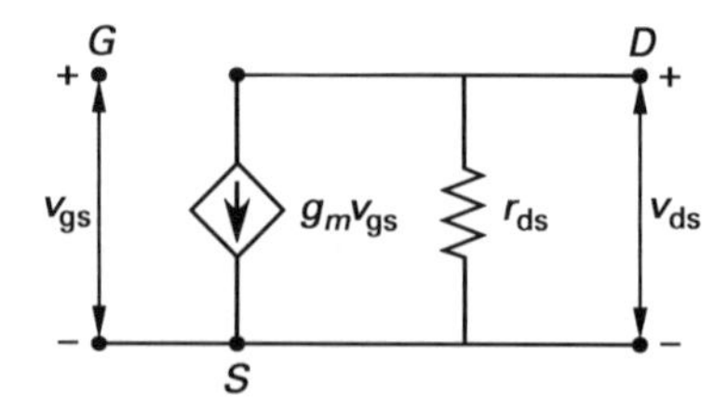

5. MOSFET CHARACTERISTICS

(See Fig. 47.6.)

6. MOSFET BIASING

Because the gate current is zero, the voltage divider is unloaded.

$$V_G = V_{DD}\left(\frac{R_2}{R_1 + R_2}\right) = V_{GS} + I_S R_S \qquad 47.7$$

The load line equation is found from Kirchhoff's voltage law and is the same as for the JFET.

$$V_{DD} = I_S(R_D + R_S) + V_{DS} \qquad 47.8$$

(See Fig. 47.7.)

Figure 47.6 *Metal-Oxide Semiconductor Field-Effect Transistor*

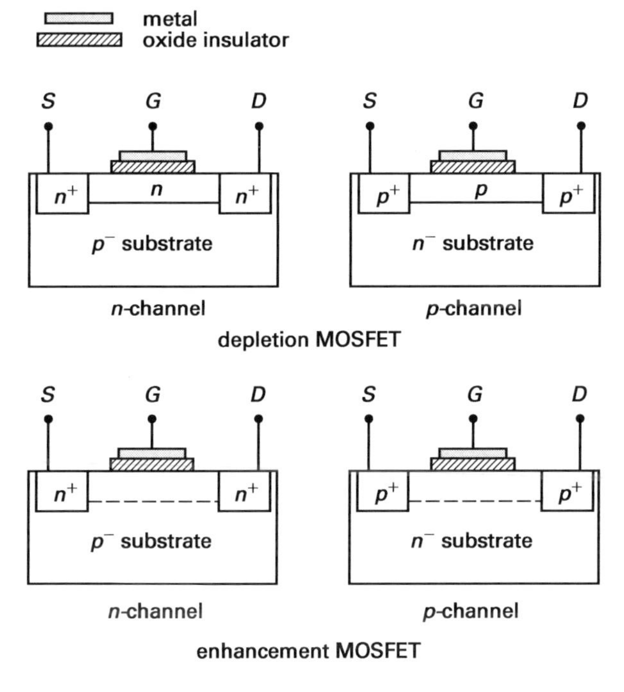

(a) conceptual construction

resistive (ohmic) region | pinchoff region | breakdown region

I_D (mA), enhancement region, I_{DSS}, depletion region, $V_G = +6, +4, +2, 0, -2, -4, -6$, V_{P0}, V_D (V)

(*n*-channel: $V_G = V_{GS}$ and $V_D = V_{DS}$)

(*p*-channel: $V_G = V_{SG}$ and $V_D = V_{SD}$)

(b) depletion-mode characteristics

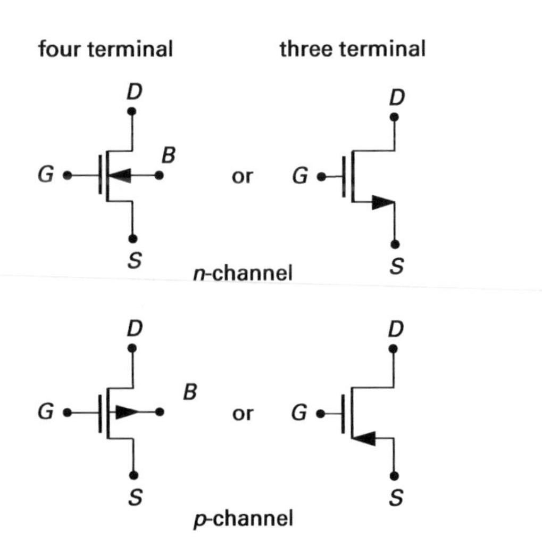

(c) depletion MOSFET symbols

four terminal | three terminal

n-channel

p-channel

(d) enhancement MOSFET symbols

Figure 47.7 Typical MOSFET Biasing Circuit

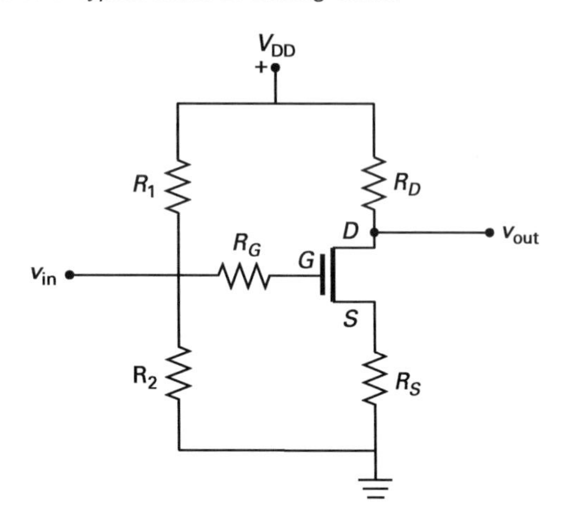

EPRM Chapter 48

Electrical and Electronic Devices

PART 2. AMPLIFIERS

1. FUNDAMENTALS

(See Fig. 48.1, Fig. 48.2, Fig. 48.4, and Fig. 48.6.)

$$v_{\text{out}} = A_V(v^+ - v^-) \quad 48.3$$

- *input signal range with the distortion restriction*

$$|v^+ - v^-| < \frac{V_{\text{DC}} - 3\text{ V}}{A_V} \quad 48.4$$

In practice, both the *differential-mode signal*, v_{dm}, and the *common-mode signal*, v_{cm}, are amplified.

$$v_{\text{dm}} = v^+ - v^- \quad 48.5$$

$$v_{\text{cm}} = \tfrac{1}{2}(v^+ + v^-) \quad 48.6$$

- *common-mode rejection ratio* (CMRR)

$$\text{CMRR} = \left|\frac{A_{\text{dm}}}{A_{\text{cm}}}\right| \quad 48.7$$

$$v_{\text{out}} = A_{\text{dm}}v_{\text{dm}}\left(1 + \left(\frac{1}{\text{CMRR}}\right)\left(\frac{v_{\text{cm}}}{v_{\text{dm}}}\right)\right) \quad 48.8$$

2. IDEAL OPERATIONAL AMPLIFIERS

- $Z_{\text{in}} = \infty$
- $Z_{\text{out}} = 0$
- $A_V = \infty$
- $\text{BW} = \infty$

Figure 48.1 Bandwidth

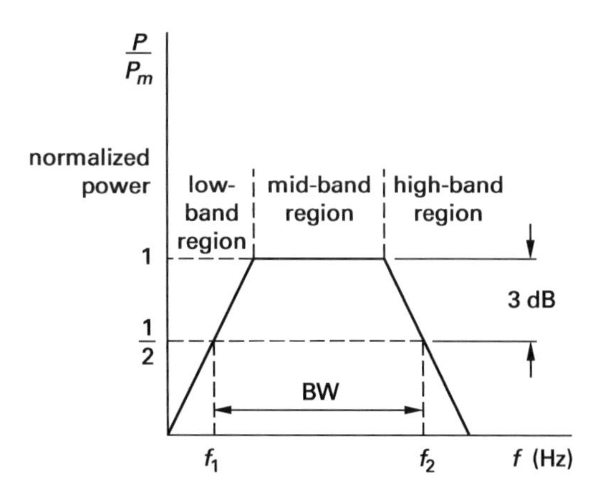

Figure 48.2 Operational Amplifier Symbols

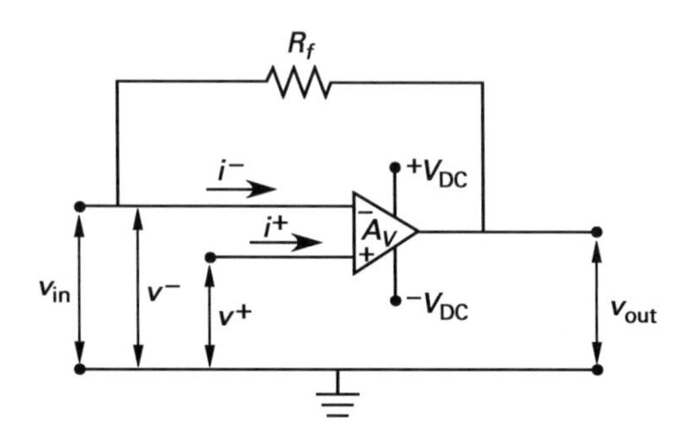

Figure 48.4 Differential Amplifier

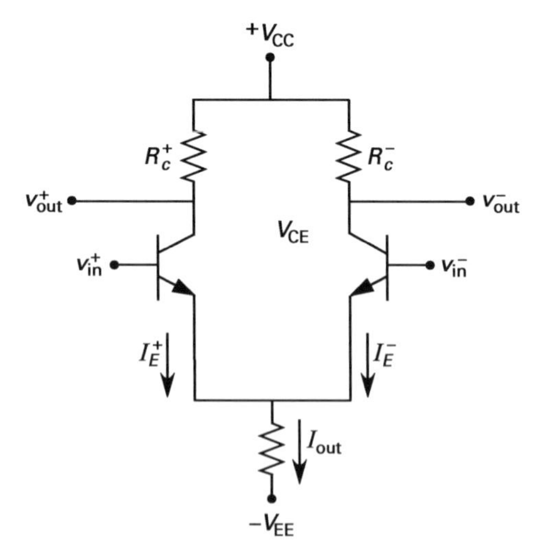

The assumptions regarding the properties of the ideal op amp result in the following practical results during analysis.

- The current to each input is zero.
- The voltage between the two input terminals is zero.
- The op amp is operating in the linear range.

The voltage difference of zero between the two terminals is called a *virtual short circuit*, or, because the positive terminal is often grounded, a *virtual ground.*

Figure 48.6 Operational Amplifier Equivalent Circuit

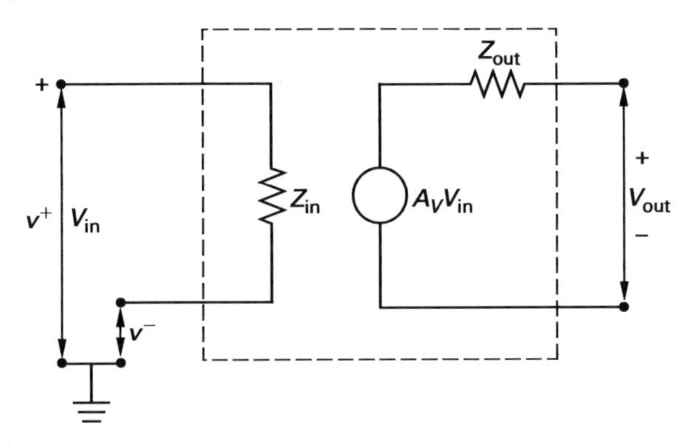

3. OPERATIONAL AMPLIFIER LIMITS

- *unity gain*

$$|A| = A_0 \frac{\omega_0}{\omega_u} = 1 \qquad 48.9$$

- *slew rate,* S_R

$$S_R = \frac{dv}{dt}\bigg|_{\max} \approx \frac{I_{\max}}{C} \qquad 48.10$$

Figure 48.9 Operational Amplifier Frequency Response

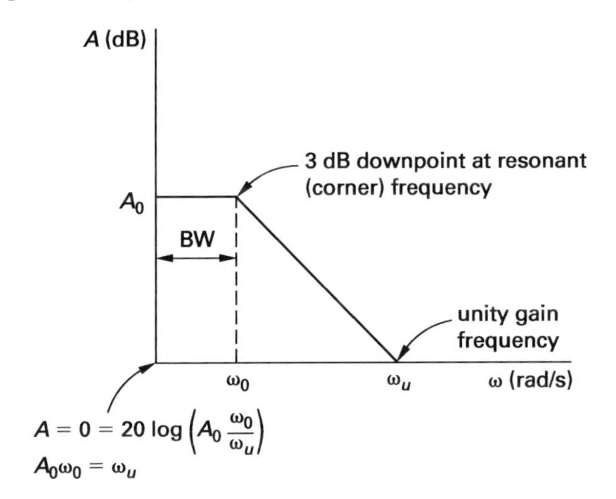

4. AMPLIFIER NOISE

- *thermal noise,* P_n (also called the *noise power*)

$$P_n = \frac{V^2_{\text{noise,rms}}}{4R} = \kappa T(\text{BW}) \qquad 48.11$$

$$\text{SNR} = 10 \log \frac{P_s}{P_n} = 20 \log \frac{V_s}{V_n} \qquad 48.12$$

PART 3. PULSE CIRCUITS: WAVEFORM SHAPING AND LOGIC

6. CLAMPING CIRCUITS

Analysis of a limiter circuit yields Eq. 48.13 and Eq. 48.14.

- *forward-biased case*

$$v_{\text{out}} = \frac{R(V_{\text{ref}} - V_F)}{R + R_f} + \frac{R_f V_{\text{in}}}{R_f + R} \quad [V_{\text{in}} < V_{\text{ref}} - V_F] \qquad 48.13$$

- *reverse-biased case* (ignoring the small reverse saturation current)

$$V_{\text{out}} = V_{\text{in}} \quad [V_{\text{in}} > V_{\text{ref}} - V_F] \qquad 48.14$$

Equation 48.13 represents the forward-biased case. Equation 48.14 represents the reverse-biased case, ignoring the small reverse saturation current.

Analysis of a clipper circuit yields Eq. 48.15 and Eq. 48.16.

$$v_{\text{out}} = \frac{R_f V_{\text{in}}}{R_f + R} + \frac{R(V_{\text{ref}} + V_F)}{R + R_f} \quad [V_{\text{in}} > V_{\text{ref}} + V_F] \qquad 48.15$$

$$V_{\text{out}} = V_{\text{in}} \quad [V_{\text{in}} < V_{\text{ref}} + V_F] \qquad 48.16$$

- *approximate transfer equation for a precision diode*

$$V_{\text{out}} = V_{\text{in}} - \frac{V_F}{A_V} \qquad 48.17$$

8. ZENER VOLTAGE REGULATOR CIRCUIT: PRACTICAL

(See Fig. 48.15.)

- *defining equations for the equivalent circuit*

$$V_{\text{in}} - V_L = (I_L + I_Z)R_s \qquad 48.24$$

$$V_L = V_{\text{ZM}} + I_Z R_Z \qquad 48.25$$

- *source resistance* (with the input voltage at its minimum value, and the load current at its maximum value)

$$V_L = \frac{V_{\text{ZM}}R_s + V_{\text{in}}R_Z}{R_s + R_Z} - I_L\left(\frac{R_s R_Z}{R_s + R_Z}\right) \qquad 48.26$$

$$I_{Z,\max} = \frac{V_{\text{ZM}}R_s + V_{\text{in,max}}R_Z}{R_Z(R_s + R_Z)} - \frac{V_{\text{ZM}}}{R_Z}$$

$$= \frac{V_{\text{in,max}}}{R_s} - \frac{V_{\text{ZM}}R_s + V_{\text{in,max}}R_Z}{R_s(R_s + R_Z)} \qquad 48.27$$

- *power requirements for the diode and the supply resistor*

$$P_D = I_{Z,\max}V_{\text{ZM}} + I^2_{Z,\max}R_Z \qquad 48.28$$

$$P_{R_s} = I^2_{Z,\max}R_s \qquad 48.29$$

Figure 48.15 Equivalent Zener Diode Voltage Regulator

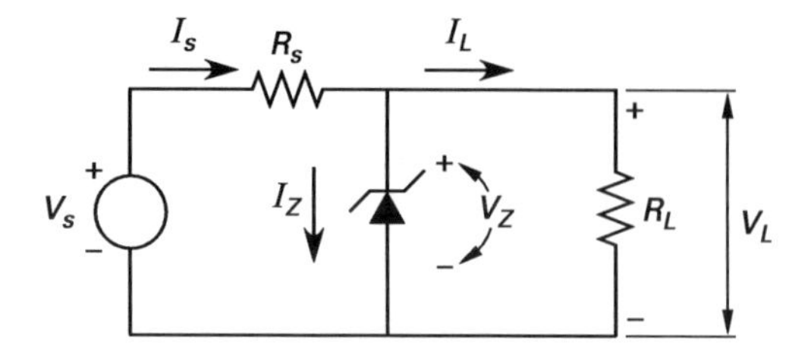

(a) regulator circuit

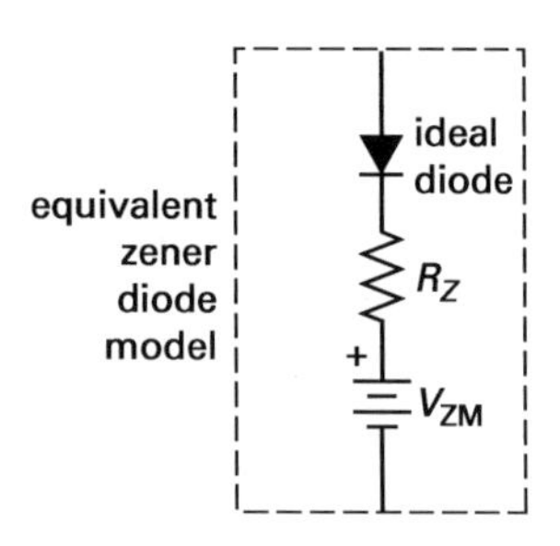

(b) regulator circuit with equivalent zener model

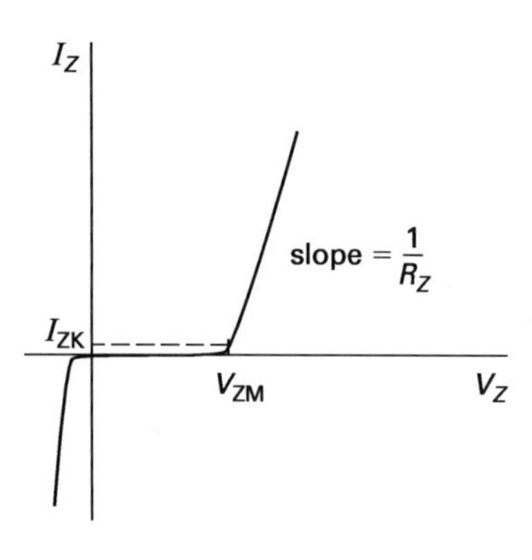

(c) characteristics

Figure 48.17 Transistor Switch

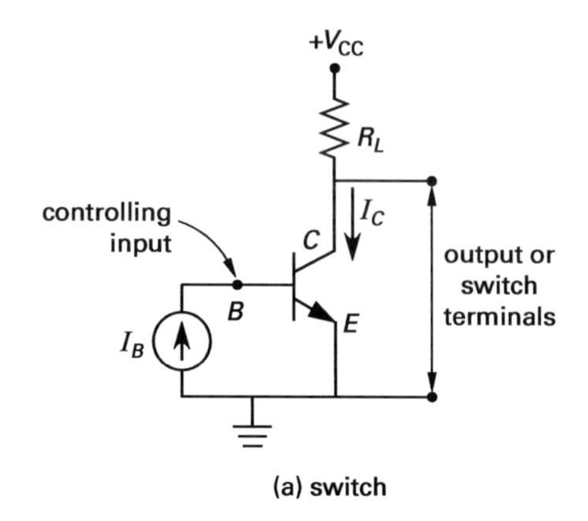

(a) switch

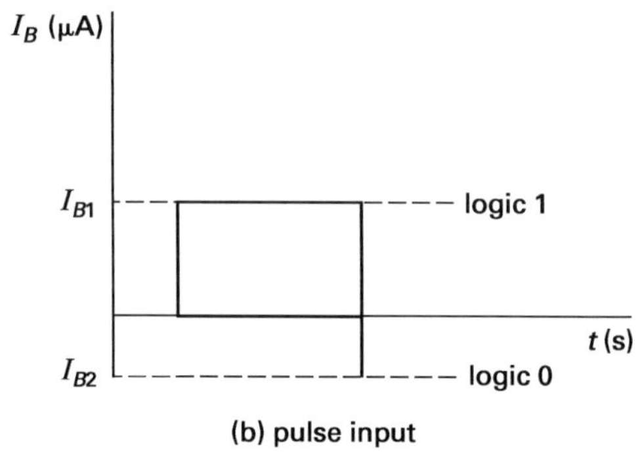

(b) pulse input

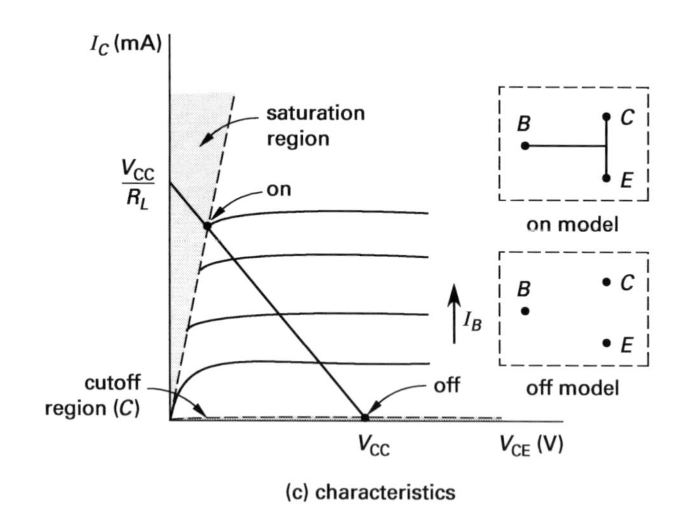

(c) characteristics

10. TRANSISTOR SWITCH FUNDAMENTALS

When the transistor switch is in the on state, that is, saturated,

$$I_{C,\text{sat}} = \frac{V_{\text{CC}} - V_{\text{CE,sat}}}{R_L} \quad 48.36$$

The collector-emitter voltage, V_{CE}, is typically 0.1 V for germanium diodes and 0.2 V for silicon diodes and can be ignored for first-order calculations—hence the short-circuit model of Fig. 48.17(c). The value of the load resistance must be determined so that when in saturation the condition of Eq. 48.37 is satisfied.

$$I_{B1} \geq \frac{I_{C,\text{sat}}}{h_{\text{FE}}} \approx \frac{V_{\text{CC}}}{h_{\text{FE}} R_L} \quad 48.37$$

11. JFET SWITCHES

(See Fig. 48.18.)

12. CMOS SWITCHES

(See Fig. 48.19 and Fig. 48.20.)

15. LOGIC GATES

(See Table 48.1.)

Figure 48.18 JFET Switch

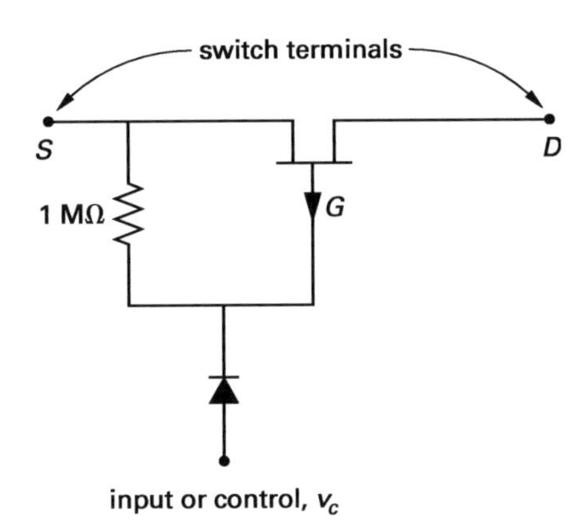

Figure 48.19 CMOS Inverter

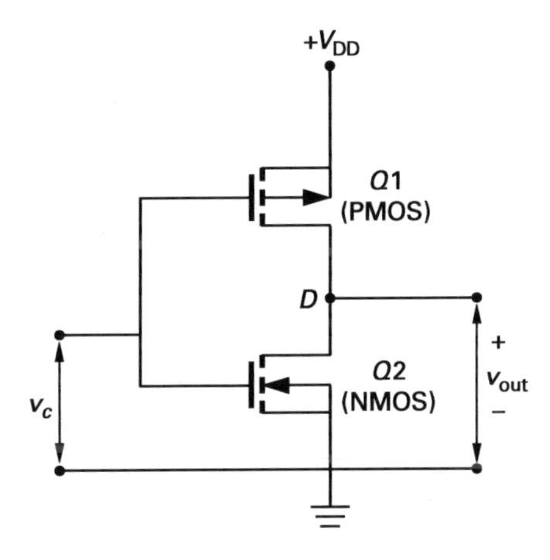

(a) digital switch

v_{out}

V_{DD}

$\approx \frac{1}{2} V_{DD}$

v_c

(b) characteristics

Figure 48.20 CMOS Switch

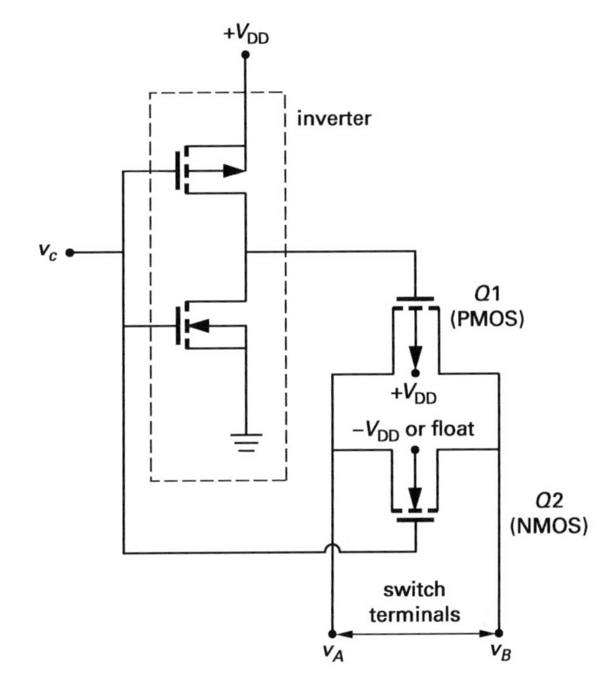

16. SIMPLIFICATION OF BINARY VARIABLES

- *commutative*

$$A + B = B + A$$
$$A \cdot B = B \cdot A$$

- *associative*

$$A + (B + C) = (A + B) + C$$
$$A \cdot (B \cdot C) = (A \cdot B) \cdot C$$

- *distributive*

$$A \cdot (B + C) = (A \cdot B) + (A \cdot C)$$
$$A + (B \cdot C) = (A + B) \cdot (A + C)$$

- *absorptive*

$$A + (A \cdot B) = A$$
$$A \cdot (A + B) = A$$

- *de Morgan's theorems*

$$\overline{A + B} = \overline{A} \cdot \overline{B}$$
$$\overline{A \cdot B} = \overline{A} + \overline{B}$$

17. LOGIC CIRCUIT FAN-OUT

(See Fig. 48.22.)

Fan-out is the number of parallel loads that can be driven from one output node of a logic circuit.

- *source fan-out*

$$N + \frac{R_b}{R_c} \leq \frac{V_{CC} - V_{BE} - I_{CBO}R_c}{I_{B,\min}R_c} \qquad 48.45$$

- *sink fan-out*

$$N \leq \frac{I_{C,\text{sat}}R_c + V_{CE,\text{sat}} - V_{CC}}{I_{CBO}R_c} \qquad 48.46$$

24. CMOS LOGIC

(See Fig. 48.31.)

25. MULTIVIBRATORS

(See Fig. 48.32 and Fig. 48.33.)

Table 48.1 *Logic Gates*

inputs		not	and	or	nand	nor	exclusive or
A	B	$-A$ or $\overline{A}$	AB	$A+B$	$\overline{AB}$	$\overline{A+B}$	$A \oplus B$
0	0	1	0	0	1	1	0
0	1	1	0	1	1	0	1
1	0	0	0	1	1	0	1
1	1	0	1	1	0	0	0

Figure 48.22 *Fan-Out Unit Loads*

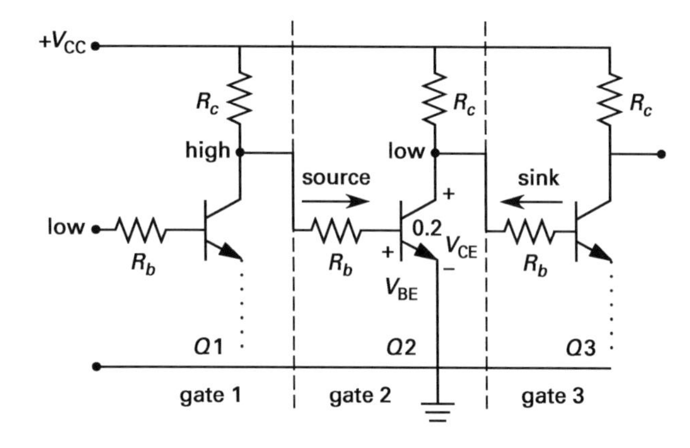

PART 4. CIRCUITS AND DEVICES

26. CIRCUIT: PHASE-LOCKED LOOP (PLL)

(See Fig. 48.34.)

EPRM Chapter 49 Digital Logic

4. FUNDAMENTAL LOGIC OPERATIONS

(See Table 49.2 and Fig. 49.2.)

6. MINTERMS AND MAXTERMS

(See Table 49.4.)

7. CANONICAL REPRESENTATION OF LOGIC FUNCTIONS

- *minterm form* (of the three-variable function in Table 49.4)

$$F(A,B,C) = \sum_{i=0}^{7} m_i \qquad 49.7$$

- *canonical sum-of-product form* (SOP)

$$\begin{aligned} F(A,B,C) &= m_0 + m_1 + m_2 + m_3 + m_4 + m_5 \\ &\quad + m_6 + m_7 \\ &= \overline{A}\,\overline{B}\,\overline{C} + \overline{A}\,\overline{B}\,C + \overline{A}\,B\,\overline{C} + \overline{A}\,B\,C \\ &\quad + A\,\overline{B}\,\overline{C} + A\,\overline{B}\,C + A\,B\,\overline{C} + A\,B\,C \end{aligned} \qquad 49.8$$

- *maxterm form* (of the three-variable function in Table 49.4)

$$F(A,B,C) = \prod_{i=0}^{7} M_i \qquad 49.9$$

- *canonical product-of-sum form* (POS)

$$\begin{aligned} F(A,B,C) &= M_0 M_1 M_2 M_3 M_4 M_5 M_6 M_7 \\ &= (A+B+C)(A+B+\overline{C}) \\ &\quad \times (A+\overline{B}+C)(A+\overline{B}+\overline{C}) \\ &\quad \times (\overline{A}+B+C)(\overline{A}+B+\overline{C}) \\ &\quad \times (\overline{A}+\overline{B}+C)(\overline{A}+\overline{B}+\overline{C}) \end{aligned} \qquad 49.10$$

Figure 48.31 *CMOS Logic*

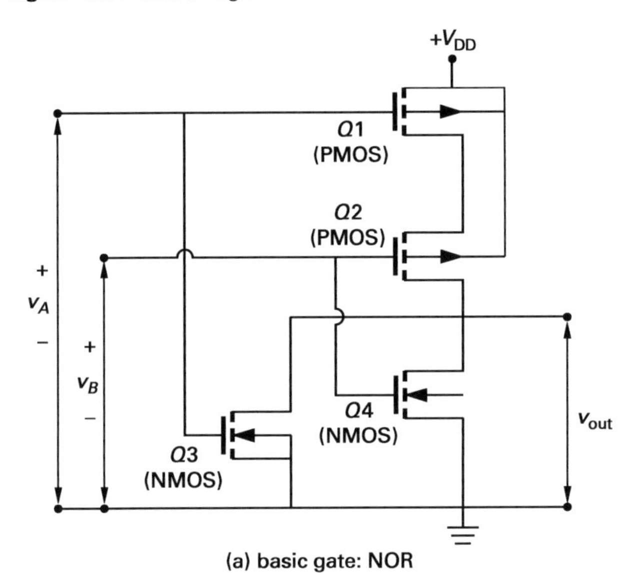

(a) basic gate: NOR

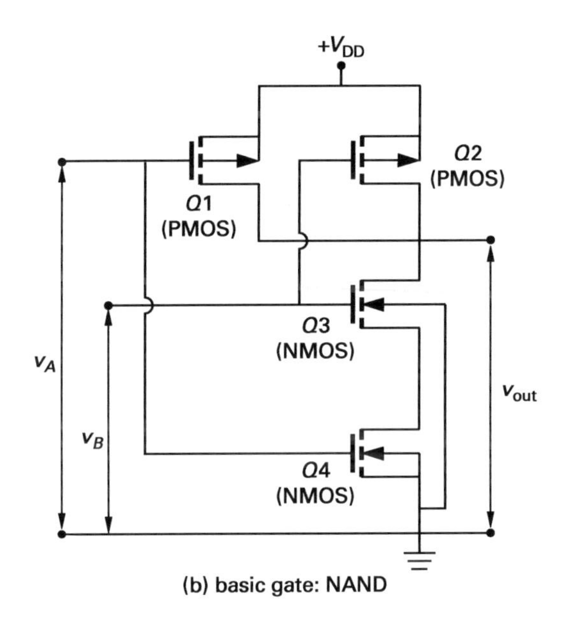

(b) basic gate: NAND

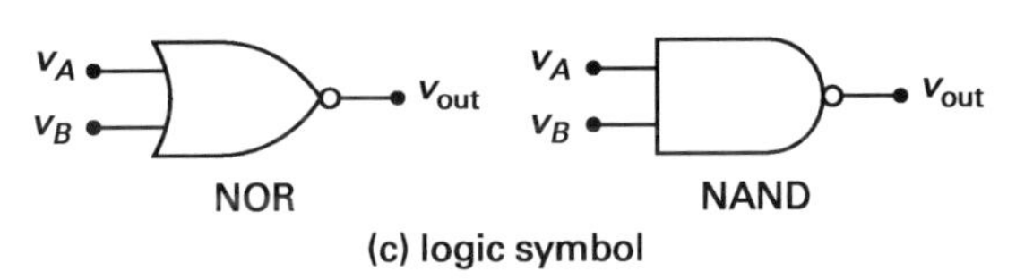

(c) logic symbol

v_A	v_B	v_{out}
L	L	H
L	H	L
H	L	L
H	H	L

NOR

v_A	v_B	v_{out}
L	L	H
L	H	H
H	L	H
H	H	L

NAND

(d) voltage truth tables

Figure 48.32 *Multivibrator Types*

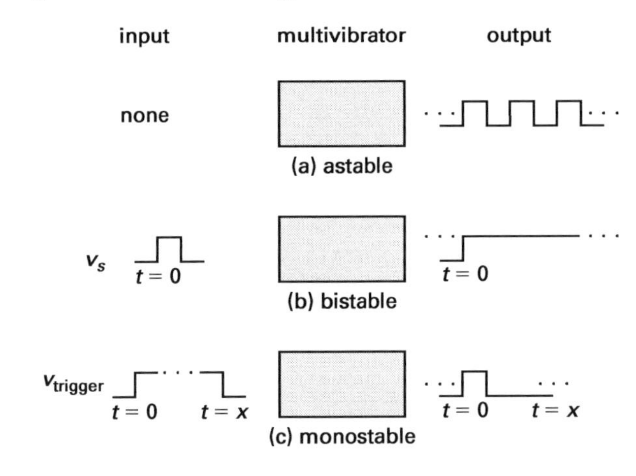

Figure 48.33 *Multivibrator [Op Amp] Circuits*

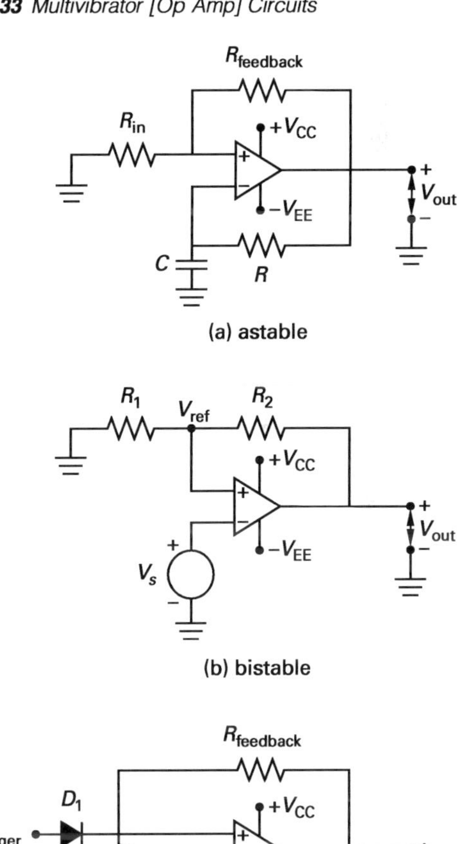

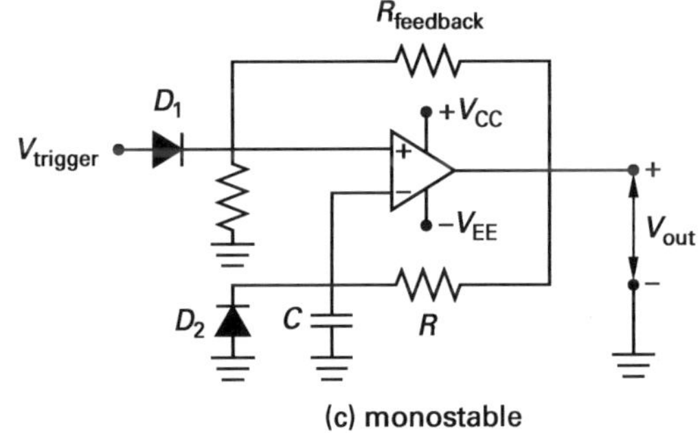

(c) monostable

Figure 48.34 *Phase-Locked Loop Functional Block Diagram*

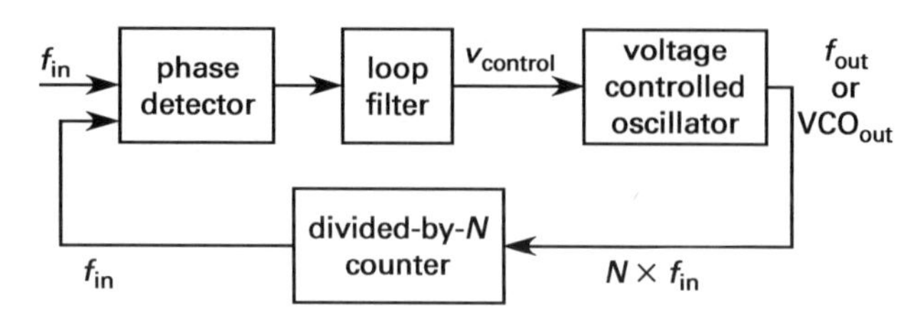

Table 49.2 *Logic Operators*

operator	symbol	A	B	C	equation
AND		0	0	0	$A \cdot B = C$
		0	1	0	
		1	0	0	
		1	1	1	
OR		0	0	0	$A + B = C$
		0	1	1	
		1	0	1	
		1	1	1	
XOR		0	0	0	$A \oplus B = C$
		0	1	1	
		1	0	1	
		1	1	0	
NAND		0	0	1	$\overline{A \cdot B} = C$
		0	1	1	
		1	0	1	
		1	1	0	
NOR		0	0	1	$\overline{A + B} = C$
		0	1	0	
		1	0	0	
		1	1	0	
XNOR or coincidence		0	0	1	$A \odot B = C$ or $A \otimes B = C$
		0	1	0	
		1	0	0	
		1	1	1	

Figure 49.2 *NOT Logic*

input variable A — B output variable

(a) symbol

A	B
0	1
1	0

(b) truth table

$$B = \overline{A}$$

(c) equation

Table 49.4 *Minterms and Maxterms*

decimal row number	binary input combinations ABC	minterm (product term)	maxterm (sum term)
0	000	$m_0 = \overline{A}\,\overline{B}\,\overline{C}$	$M_0 = A + B + C$
1	001	$m_1 = \overline{A}\,\overline{B}C$	$M_1 = A + B + \overline{C}$
2	010	$m_2 = \overline{A}B\overline{C}$	$M_2 = A + \overline{B} + C$
3	011	$m_3 = \overline{A}BC$	$M_3 = A + \overline{B} + \overline{C}$
4	100	$m_4 = A\overline{B}\,\overline{C}$	$M_4 = \overline{A} + B + C$
5	101	$m_5 = A\overline{B}C$	$M_5 = \overline{A} + B + \overline{C}$
6	110	$m_6 = AB\overline{C}$	$M_6 = \overline{A} + \overline{B} + C$
7	111	$m_7 = ABC$	$M_7 = \overline{A} + \overline{B} + \overline{C}$

Special Applications

EPRM Chapter 50
Lightning Protection and Grounding

2. CONCEPTS AND DEFINITIONS

Figure 50.2 *Standard Lightning Impulse*

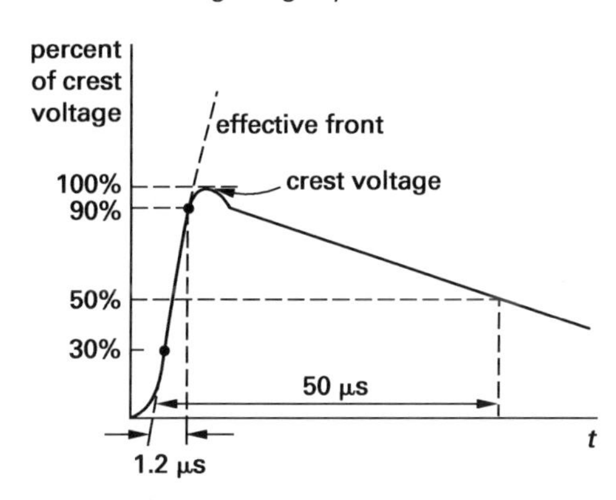

4. GROUNDING MODELS

$$\frac{l}{\delta} = \frac{l}{\frac{1}{\sqrt{\pi f \mu \sigma}}} \begin{cases} < 0.1 \text{ DC analysis} \\ > 0.1 \text{ electromagnetic analysis} \end{cases} \qquad 50.1$$

5. PROTECTIVE DEVICES

When overvoltages occur, the transient is minimized and the system is protected by connecting the overvoltage condition to ground until the energy is dissipated. This connection must be temporary, and ideally will not allow current flow at normal operating voltages. This is accomplished by the use of strategically placed *arresters*, also called *surge arresters.*

- *protection quality index* (PQI)

$$\text{PQI} = \frac{V_r}{V_p} \qquad 50.2$$

- *protective margin* (PM)

$$\text{PM} = \frac{V_w - V_p}{V_p} \qquad 50.3$$

EPRM Chapter 51
Illumination

1. HISTORY AND OVERVIEW

Every object of mass has an associated wavelength given by

$$\lambda = \frac{h}{m\text{v}} \qquad 51.1$$

2. ELECTROMAGNETIC WAVES

Figure 51.1 *Electromagnetic Spectrum: Ultraviolet, Visible, and Infrared Radiant Energy*

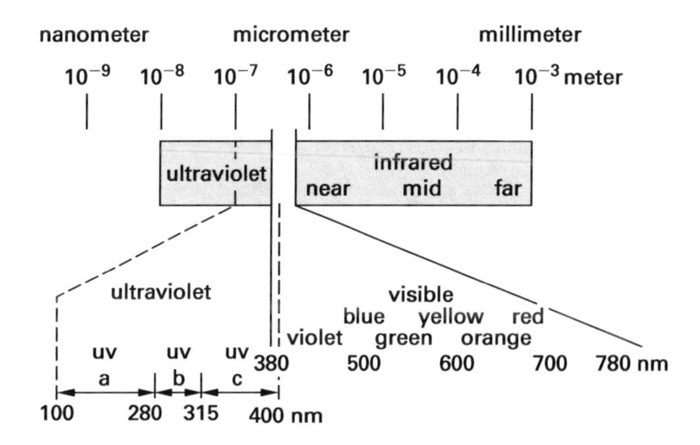

Equation 51.2 is valid for all types of electromagnetic radiation.

$$c = \lambda \nu \qquad 51.2$$

- *velocity*, v, and the *index of refraction*, n (also called the *refractive index*)

$$\lambda = \frac{\text{v}n}{\nu} \qquad 51.5$$

Wavelength bends the light (electromagnetic wave) toward the surface normal in a material denser than the original medium, and away from the normal in a less dense medium.

$$E = h\nu = \frac{hc}{\lambda} = \frac{h\text{v}n}{\lambda} \qquad 51.6$$

4. ELECTROMAGNETIC SPECTRUM: VISIBLE LIGHT

Table 51.3 *Color Versus Wavelength*

color	wavelength (nm)
violet	380–450
blue	450–495
green	495–570
yellow	570–590
orange	590–620
red	620–750

6. BLACKBODY RADIATION

Thermal radiation is exchanged according to Eq. 51.7. α is the spectral absorption factor; ρ is the spectral reflection factor; and τ is the spectral transmission factor. All these elements are dependent on the wavelength, λ.

$$\alpha + \rho + \tau = 1 \qquad 51.7$$

7. PLANCK RADIATION LAW

The *Planck radiation law* is an expression providing the spectral radiance, L, of a blackbody as a function of wavelength and temperature. Planck's law provided the shape of the curves in Fig. 51.3.

- *a form of the law* (in terms of wavelength)

$$L(\lambda,T) = L_\lambda = \left(\frac{2hc^2}{\lambda^5}\right)\left(\frac{1}{e^{hc/\lambda\kappa T}-1}\right) \qquad 51.9$$

$$L(f,T) = L_f = \left(\frac{2hf^3}{c^2}\right)\left(\frac{1}{e^{hf/\kappa T}-1}\right) \qquad 51.10$$

$$L(\lambda,T)d\lambda = L(f,T)df \qquad 51.11$$

- *Wien radiation law* (applicable to the shaded region in Fig. 51.4 between 2000K and 3400K)

$$L(\lambda,T) = L_\lambda = \left(\frac{2hc^2}{\lambda^5}\right)e^{-hc/\lambda\kappa T} \qquad 51.12$$

Figure 51.3 *Radiated Energy Versus Wavelength as Temperature Varies*

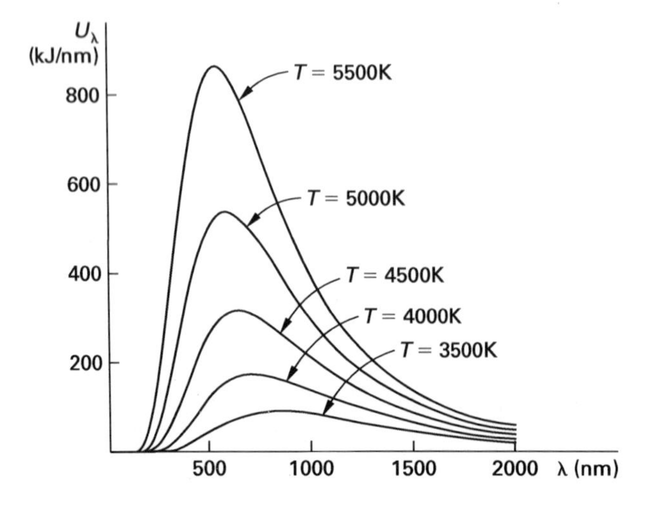

8. WIEN DISPLACEMENT LAW

The peaks calculated from Eq. 51.13 and Eq. 51.14 are located on the dashed line between points A and B in Fig. 51.4.

$$\lambda_{max} = \frac{b}{T} \qquad 51.13$$

b' is a constant, 4.0956×10^{-4} W·sr·m^2·m·K^5.

$$L_{max} = b'T^5 \qquad 51.14$$

Figure 51.4 *Blackbody Radiation Curves*

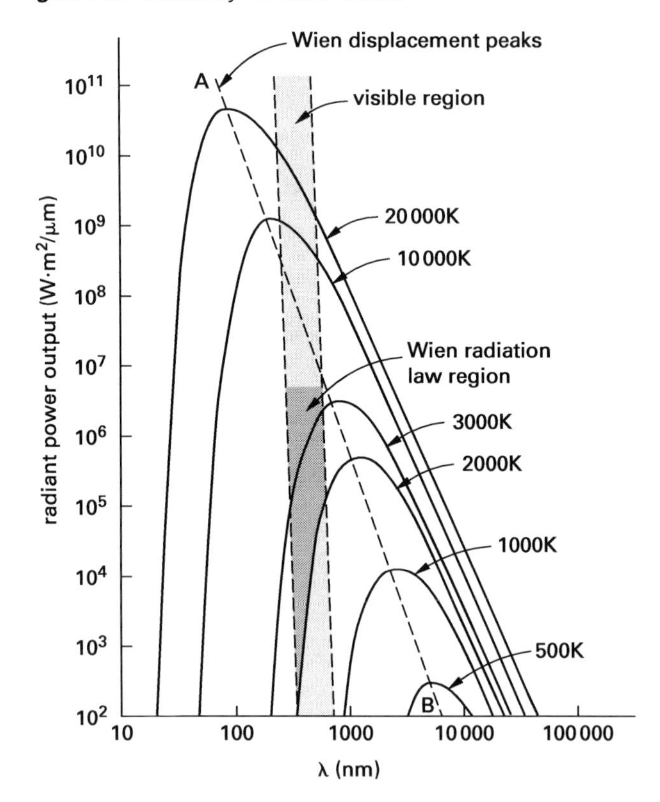

9. STEFAN-BOLTZMANN LAW

- *fourth-power law* or *Stefan's law*

$$P = \sigma A T^4 \qquad 51.15$$

- *fourth-power law* (in terms of the *luminous exitance, M*)

$$M_v = \sigma T^4 \qquad 51.16$$

- *fourth-power law* (in terms of the *radiance, L*)

$$L_e = \sigma_L T^4 \qquad 51.17$$

10. GRAYBODY AND SELECTIVE RADIATORS

$$\epsilon = \frac{M_e}{M_{\text{blackbody}}} \qquad 51.18$$

12. ATOMIC STRUCTURE: ELECTROMAGNETIC RADIATION

$$E_2 - E_1 = h\nu_{21} \qquad 51.19$$

- *wavelength* (in nanometers and in terms of the potential difference, in volts)

$$\lambda_{\text{nm}} = \frac{1239.76\ \frac{\text{nm}}{\text{V}}}{\Delta V} \qquad 51.20$$

- *wavelength* (in terms of the energy gap shown in Fig. 51.8)

$$\lambda = \frac{hc}{E_g} \qquad 51.21$$

Figure 51.8 *Light Emitting Diode (LED) p-n Junction*

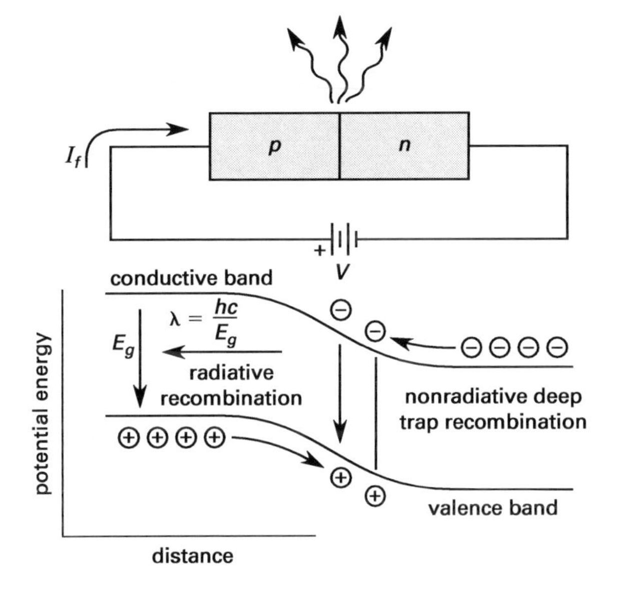

(See Fig. 51.9.)

13. THE CANDELA AND LUMEN

There is a total of 683 lm available per watt at 555 nm, though the maximum luminous efficacy (all radiation in the visible band) for an ideal white source provides only 220 lm/W.

(See Fig. 51.10 and Fig. 51.11.)

Figure 51.9 *Fluorescent Phenomena*

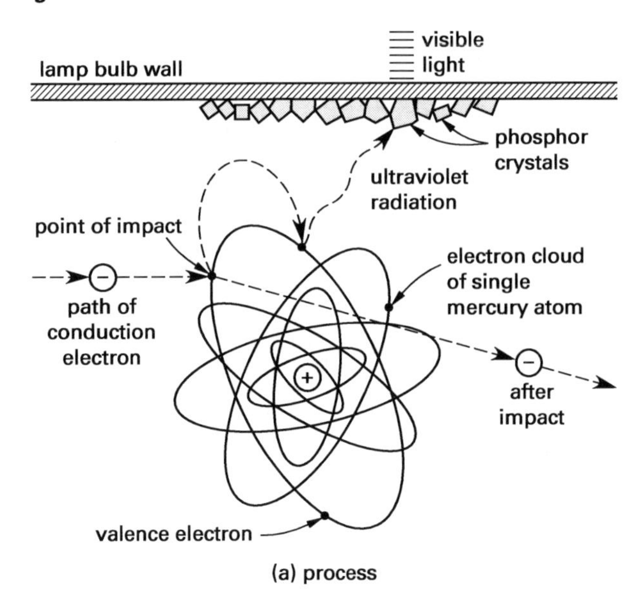

(a) process

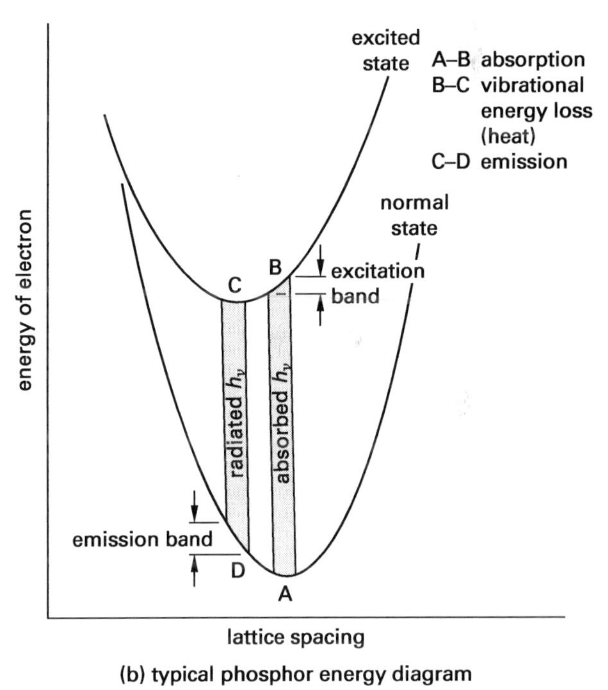

(b) typical phosphor energy diagram

15. ILLUMINANCE

The symbol changes from E to E_v for illuminance, and to E_e for irradiance.

$$E = \frac{d\Phi}{dA} \qquad 51.27$$

$$E = \frac{\Phi_t}{4\pi r^2} \qquad 51.28$$

(See Table 51.6.)

Figure 51.10 *Relationships Between Candela, Lumen, Lux, and Foot-candle*

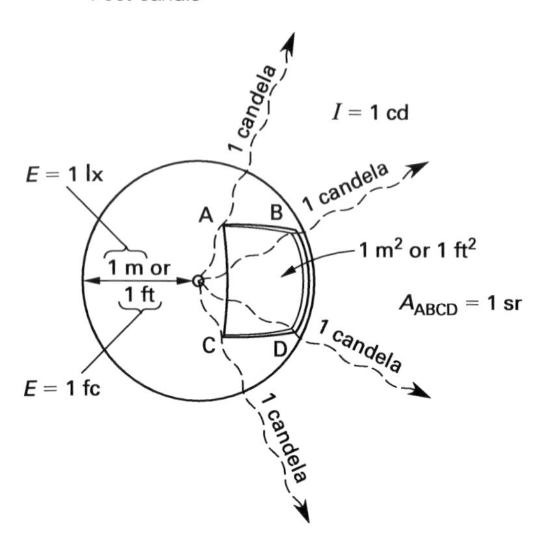

Figure 51.11 *Illumination Terms and Relationships*

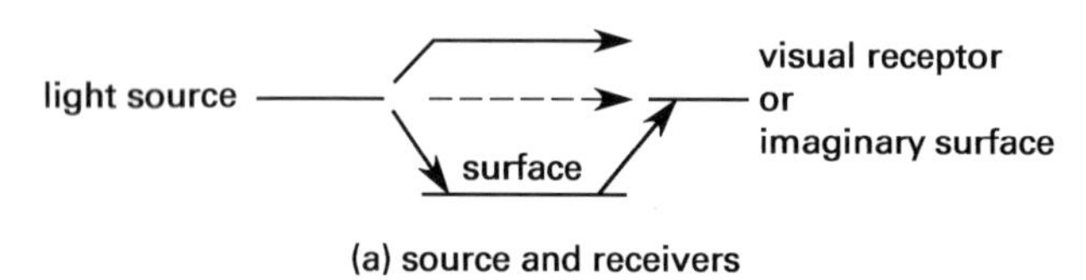

(a) source and receivers

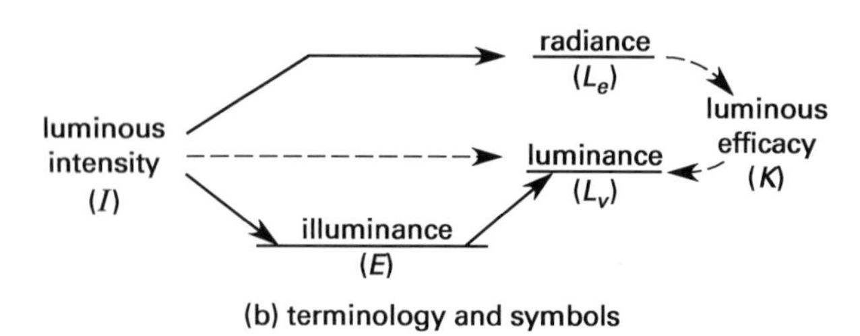

(b) terminology and symbols

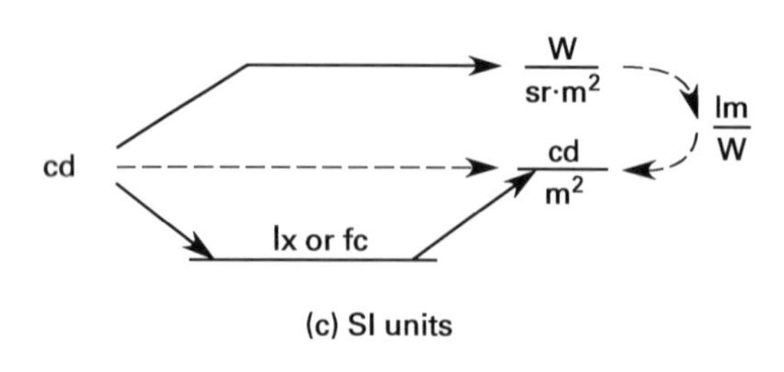

(c) SI units

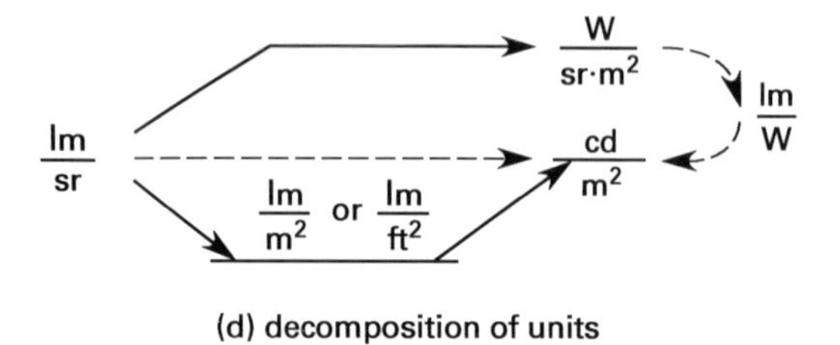

(d) decomposition of units

Table 51.6 *Recommended Illuminance*

category	location/condition/ magnitude	lux	fc
A	public spaces	30	3
B	simple orientation/short visits	50	5
C	cooking space/simple visual tasks	100	10
D	visual task/high contrast required/large size	300	30
E	visual task/high contrast/ small size	500	50
F	visual task/low contrast/ small size	1000	100
G	critical visual tasks	3000–10 000	300–1000

16. ILLUMINANCE: INVERSE SQUARE LAW

The inverse square law is accurate to within 1% when the distance d is at least five times the maximum dimension of the source (or luminaire). This is called the *five-times rule.*

$$E = \frac{I}{d^2} \qquad 51.29$$

$$E_1 r_1^2 = E_2 r_2^2 \qquad 51.30$$

(See Fig. 51.12.)

17. ILLUMINANCE: LAMBERT'S LAW

$$E_2 = E_1 \cos\theta \qquad 51.31$$

$$E = \frac{I}{d^2}\cos\theta \qquad 51.32$$

18. ILLUMINANCE: COSINE-CUBED LAW

$$E = \frac{I\cos^3\theta}{h^2} \qquad 51.33$$

20. INTERACTION OF LIGHT WITH MATTER

(See Fig. 51.14.)

Figure 51.12 *Illuminance Laws*

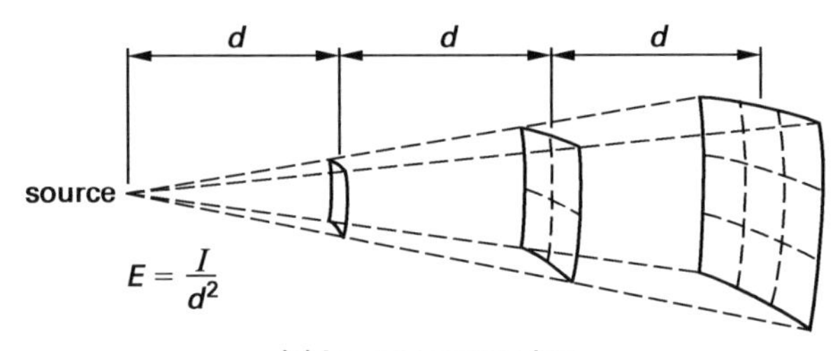

(a) inverse square law

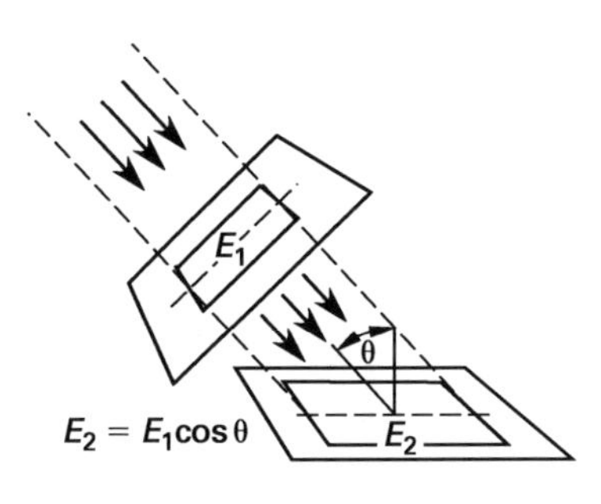

(b) Lambert's law

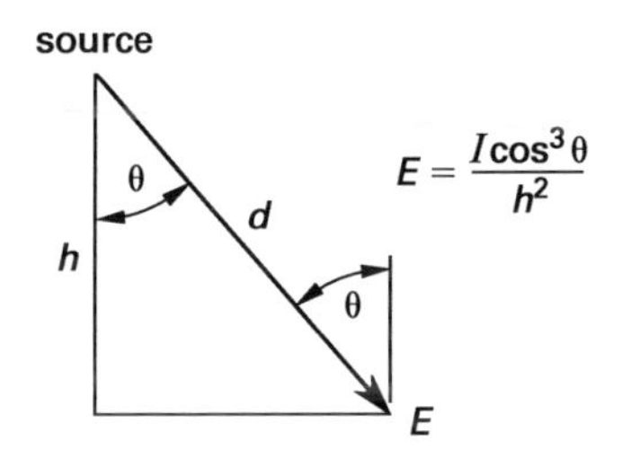

(c) cosine cubed law

21. REFLECTION

$$\sin\theta_c = \frac{1}{n} \quad 51.35$$

Figure 51.16 *Reflection from a Surface*

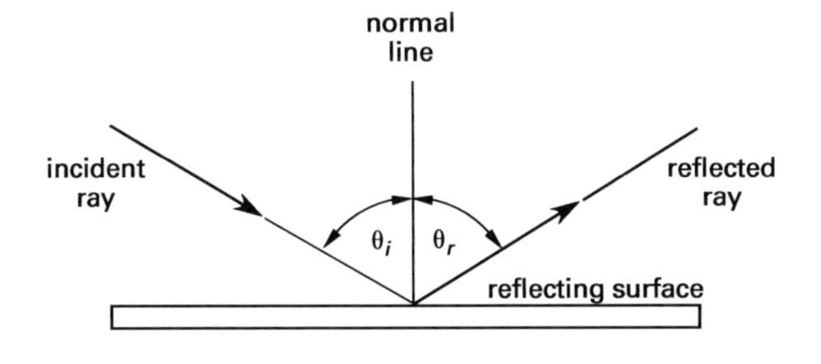

Figure 51.14 *Reflections*

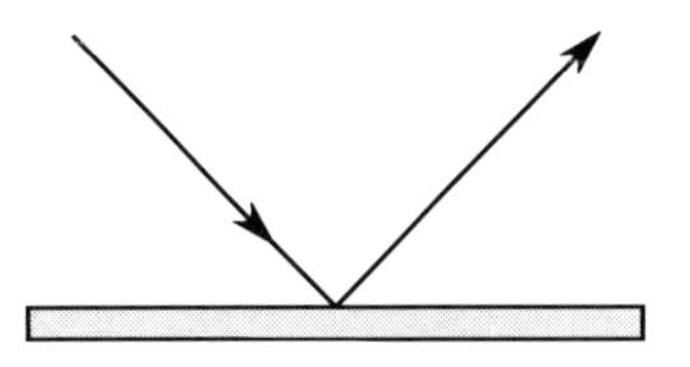

(a) polished surface, specular

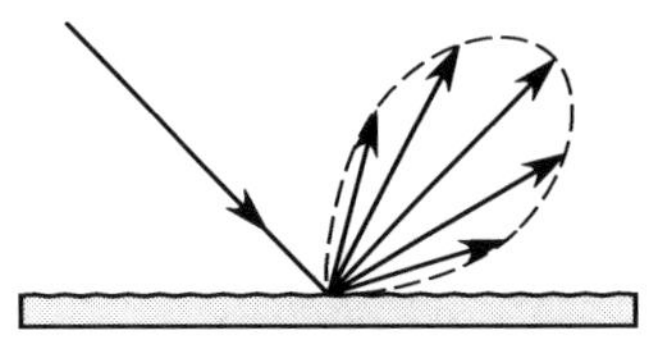

(b) rough surface, spread

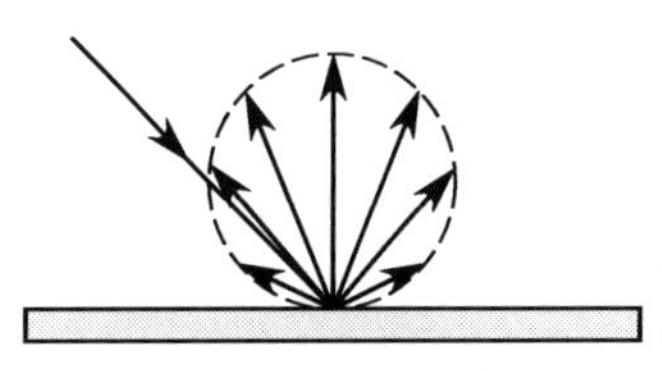

(c) matte surface, diffuse

22. REFRACTION

$$n_{\text{relative}} = \frac{n_2}{n_1} = \frac{\sin\theta_1}{\sin\theta_2} \quad 51.36$$

Figure 51.18 *Refraction of Light*

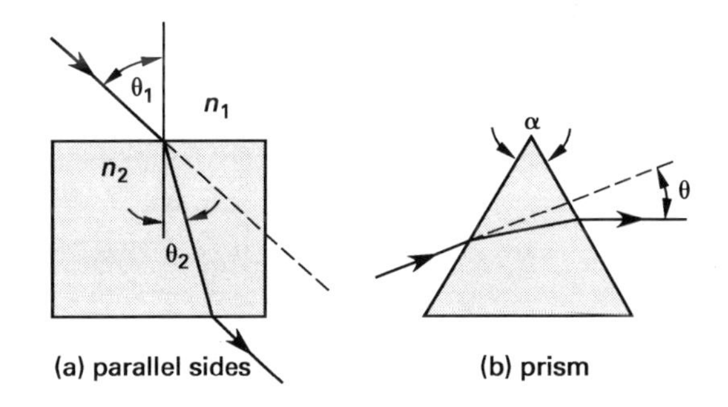

(a) parallel sides (b) prism

23. INDEX OF REFRACTION

$$n = \frac{c_{\text{vacuum}}}{c_{\text{medium}}} = \frac{3\times 10^8\ \frac{\text{m}}{\text{s}}}{c_{\text{medium,m/s}}} \quad 51.39$$

24. DIFFRACTION

$$\sin\theta = \frac{m\lambda}{d} \quad \text{[in phase]} \quad 51.40$$

$$\sin\theta = \frac{(2m+1)\lambda}{2d} \quad \text{[out of phase]} \quad 51.41$$

26. INTERFERENCE FROM SLITS

For the mth reinforcement (light band, maximum, etc.),

$$s_1 - s_2 = m\lambda \quad \text{[in phase]} \qquad 51.42$$

$$y = \frac{m\lambda x}{d} \quad \text{[in phase]} \qquad 51.43$$

For the mth cancellation (dark band, minimum, etc.),

$$s_1 - s_2 = \frac{(2m+1)\lambda}{2} \quad \text{[out of phase]} \qquad 51.44$$

$$y = \frac{(2m+1)\lambda x}{2d} \quad \text{[out of phase]} \qquad 51.45$$

27. INTERFERENCE FROM THIN FILMS

It is essential that the wavelength in the film, λ_{film}, be used, not the free space wavelength.

$$(m + {}^1/_2)\lambda_{\text{film}} = 2t \quad \text{[in phase]} \qquad 51.46$$

$$m\lambda_{\text{film}} = 2t \quad \text{[out of phase]} \qquad 51.47$$

$$\lambda_{\text{film}} = \frac{\lambda_{\text{vacuum}}}{n_{\text{film}}} \qquad 51.48$$

31. DAYLIGHT LUMEN METHOD

step 1: Calculate exterior illuminances using the CIE standard spectral radian power distributions.

step 2: Accounting for fenestration, determine the net transmittance into the room.

step 3: Apply *coefficients of utilization* (CU).

step 4: Calculate the product of the factors from steps 1, 2, and 3 to determine the interior illuminance.

$$E_i = E_{xh}\tau_{\text{net}}(\text{CU})\left(\frac{A_s}{A_w}\right) \qquad 51.49$$

- net transmittance

$$\tau_{d \text{ or } D} = (T_{d \text{ or } D})\eta_w R_a T_c(\text{LLF}) \qquad 51.50$$

32. DAYLIGHT FACTOR METHOD

The daylight factor, DF, is a ratio of the illuminance at a point to the illuminance from the unobstructed sky.

$$\text{DF} = \text{SC} + \text{ERC} + \text{IRC} \qquad 51.56$$

SC represents the sky component, ERC is the externally reflected component, and IRC is the internally reflected component.

34. LUMEN METHOD

- average initial expected illuminance

$$E_{\text{initial}} = \frac{L_{\text{total}}(\text{CU})}{A_w} \qquad 51.58$$

- average minimum lighting level

$$E_{\text{maintained}} = \frac{L_{\text{total}}(\text{CU})(\text{LLF})}{A_w} \qquad 51.59$$

- number of lighting devices

$$E_{\text{maintained}} = \frac{N_{\text{lights}}L_{\text{per light}}(\text{CU})(\text{LLF})}{A_w} \qquad 51.60$$

35. CAVITY RATIOS

Figure 51.24 Zonal Cavity Terminology

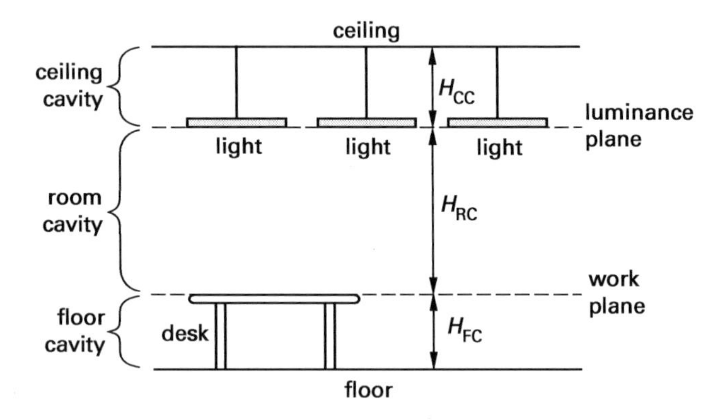

$$\text{CR} = \frac{5H_{xx}(L+W)}{LW} \qquad 51.61$$

Measurement and Instrumentation

EPRM Chapter 53 **Measurement and Instrumentation**

2. SIGNAL REPRESENTATION

$$V_{\text{ave}} = \frac{2}{\pi} V_p \qquad 53.7$$

$$V_{\text{rms}} = \frac{1}{\sqrt{2}} V_p \qquad 53.8$$

$$V_{\text{rms}} = 1.11 V_{\text{ave}} \qquad 53.9$$

3. MEASUREMENT CIRCUIT TYPES

(See Fig. 53.1.)

- *relationship between the resistors* (when balanced)

$$\frac{R_x}{R_3} = \frac{R_2}{R_4} \qquad 53.10$$

5. DC VOLTMETERS

(See Fig. 53.3.)

$$\frac{1}{I_{\text{fs}}} = \frac{R_{\text{ext}} + R_{\text{coil}}}{V_{\text{fs}}} \qquad 53.11$$

6. DC AMMETERS

(See Fig. 53.4.)

$$I_{\text{design}} = \frac{V_{\text{fs}}}{R_{\text{shunt}}} + I_{\text{fs}} \qquad 53.12$$

- Two wattmeter pf.

$$\sqrt{3}\left(\frac{P_H - P_L}{P_H + P_L}\right) = \tan^{-1}\phi$$

Figure 53.1 *Wheatstone Bridge*

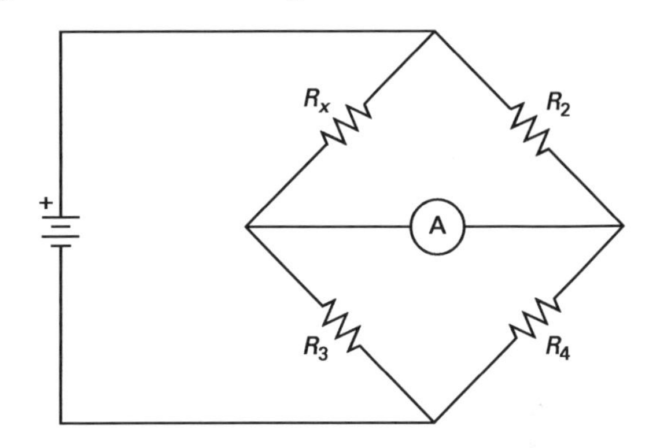

Figure 53.3 *DC Voltmeter*

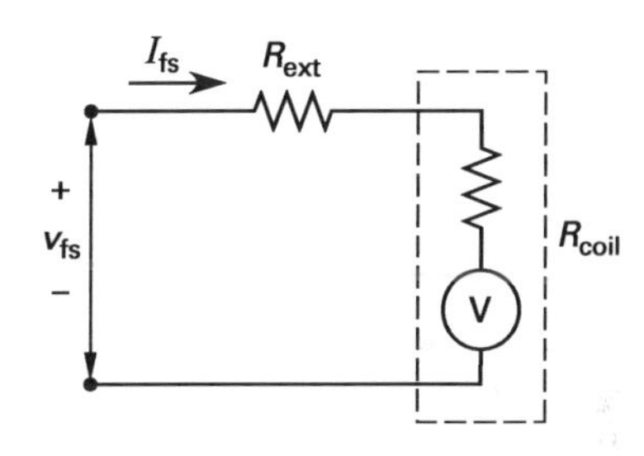

Figure 53.4 *DC Ammeter*

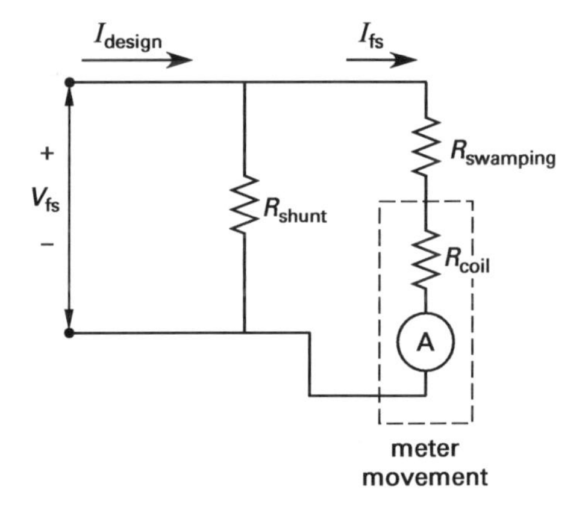

Electrical Materials

EPRM Chapter 54
Electrical Materials

1. OVERVIEW

(See the "Relative Atomic Weights" table.)

2. TYPES OF MATERIALS

Table 54.1 *Elements in Polymers*

element atomic number	element name
1	hydrogen
6	carbon
7	nitrogen
8	oxygen
9	fluorine
14	silicon

Relative Atomic Weights (referred to carbon-12)

name	symbol	atomic number	atomic weight	name	symbol	atomic number	atomic weight
actinium	Ac	89	–	meitnerium	Mt	109	–
aluminum	Al	13	26.9815	mendelevium	Md	101	–
americium	Am	95	–	mercury	Hg	80	200.59
antimony	Sb	51	121.760	molybdenum	Mo	42	95.96
argon	Ar	18	39.948	neodymium	Nd	60	144.242
arsenic	As	33	74.9216	neon	Ne	10	20.1797
astatine	At	85	–	neptunium	Np	93	237.048
barium	Ba	56	137.327	nickel	Ni	28	58.693
berkelium	Bk	97	–	niobium	Nb	41	92.906
beryllium	Be	4	9.0122	nitrogen	N	7	14.0067
bismuth	Bi	83	208.980	nobelium	No	102	–
bohrium	Bh	107	–	osmium	Os	76	190.23
boron	B	5	10.811	oxygen	O	8	15.9994
bromine	Br	35	79.904	palladium	Pd	46	106.42
cadmium	Cd	48	112.411	phosphorus	P	15	30.9738
calcium	Ca	20	40.078	platinum	Pt	78	195.084
californium	Cf	98	–	plutonium	Pu	94	–
carbon	C	6	12.0107	polonium	Po	84	–
cerium	Ce	58	140.116	potassium	K	19	39.0983
cesium	Cs	55	132.9054	praseodymium	Pr	59	140.9077
chlorine	Cl	17	35.453	promethium	Pm	61	–
chromium	Cr	24	51.996	protactinium	Pa	91	231.0359
cobalt	Co	27	58.9332	radium	Ra	88	–
copernicium	Cn	112	–	radon	Rn	86	226.025
copper	Cu	29	63.546	rhenium	Re	75	186.207
curium	Cm	96	–	rhodium	Rh	45	102.9055
darmstadtium	Ds	110	–	roentgenium	Rg	111	–
dubnium	Db	105	–	rubidium	Rb	37	85.4678
dysprosium	Dy	66	162.50	ruthenium	Ru	44	101.07
einsteinium	Es	99	–	rutherfordium	Rf	104	–
erbium	Er	68	167.259	samarium	Sm	62	150.36
europium	Eu	63	151.964	scandium	Sc	21	44.956
fermium	Fm	100	–	seaborgium	Sg	106	–
fluorine	F	9	18.9984	selenium	Se	34	78.96
francium	Fr	87	–	silicon	Si	14	28.0855
gadolinium	Gd	64	157.25	silver	Ag	47	107.868
gallium	Ga	31	69.723	sodium	Na	11	22.9898
germanium	Ge	32	72.64	strontium	Sr	38	87.62
gold	Au	79	196.9666	sulfur	S	16	32.065
hafnium	Hf	72	178.49	tantalum	Ta	73	180.94788
hassium	Hs	108	–	technetium	Tc	43	–
helium	He	2	4.0026	tellurium	Te	52	127.60
holmium	Ho	67	164.930	terbium	Tb	65	158.925
hydrogen	H	1	1.00794	thallium	Tl	81	204.383
indium	In	49	114.818	thorium	Th	90	232.038
iodine	I	53	126.90447	thulium	Tm	69	168.934
iridium	Ir	77	192.217	tin	Sn	50	118.710
iron	Fe	26	55.845	titanium	Ti	22	47.867
krypton	Kr	36	83.798	tungsten	W	74	183.84
lanthanum	La	57	138.9055	uranium	U	92	238.0289
lawrencium	Lr	103	–	vanadium	V	23	50.942
lead	Pb	82	207.2	xenon	Xe	54	131.293
lithium	Li	3	6.941	ytterbium	Yb	70	173.054
lutetium	Lu	71	174.9668	yttrium	Y	39	88.906
agnesium	Mg	12	24.305	zinc	Zn	30	65.38
ganese	Mn	25	54.9380	zirconium	Zr	40	91.224

Codes and Standards

EPRM Chapter 56
National Electrical Code

2. INTRODUCTION

Figure 56.1 *NEC Coverage*

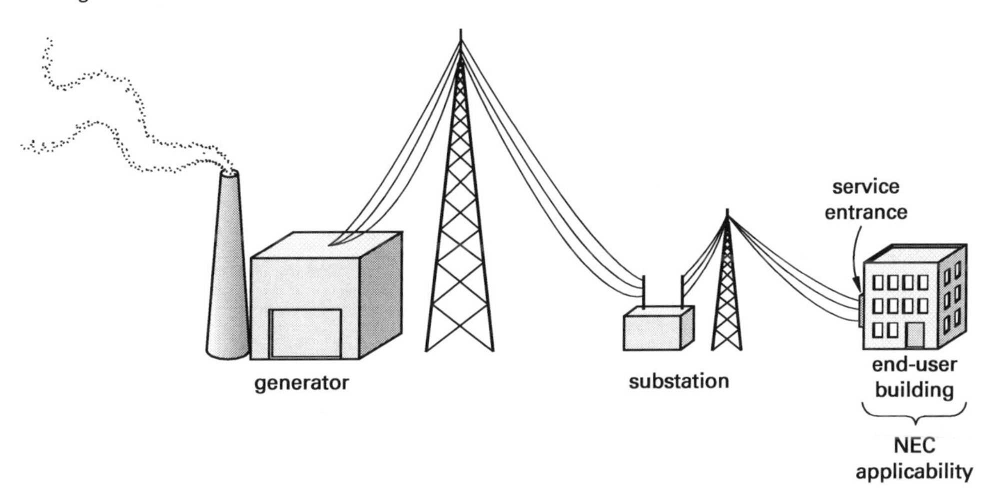

3. GENERAL

Figure 56.2 *Typical Distribution System*

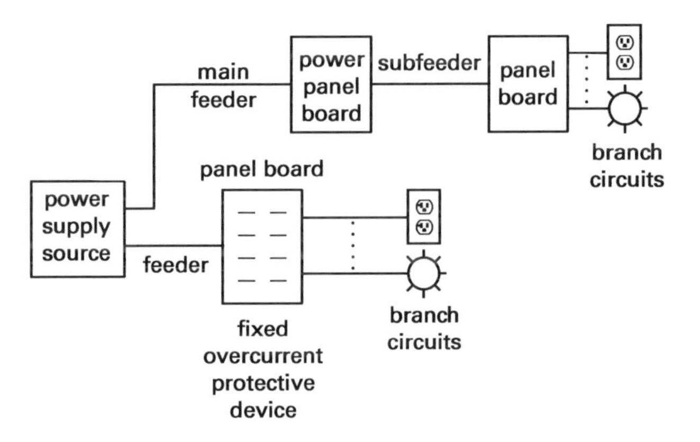

4. WIRING AND PROTECTION

(See Fig. 56.3.)

6. WIRING AND PROTECTION: BRANCH CIRCUITS

A *ground-fault circuit interrupter* (GFCI) is a device that de-energizes a circuit within an established period of time when a current to ground exceeds some predetermined value, which is less than that required to operate the overcurrent protective device of the supply circuit. The concept of such a device is illustrated in Fig. 56.4.

As long as the current is balanced in the lines running through the transformer core, no induced voltage is felt, no current flows in the tripping mechanism, and the GFCI contact remains shut.

Figure 56.3 *Four-Wire System Nomenclature*

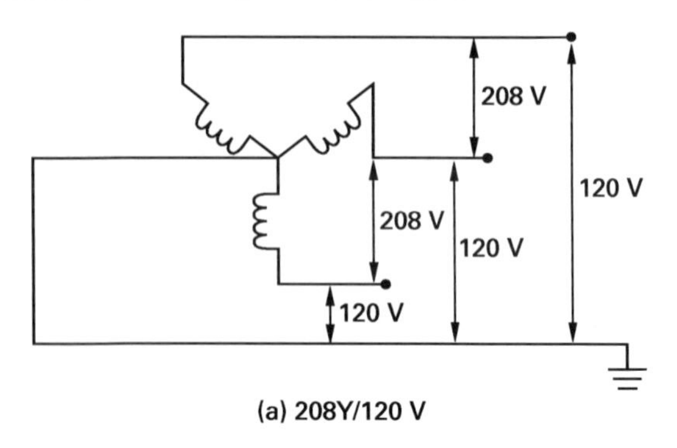

(a) 208Y/120 V

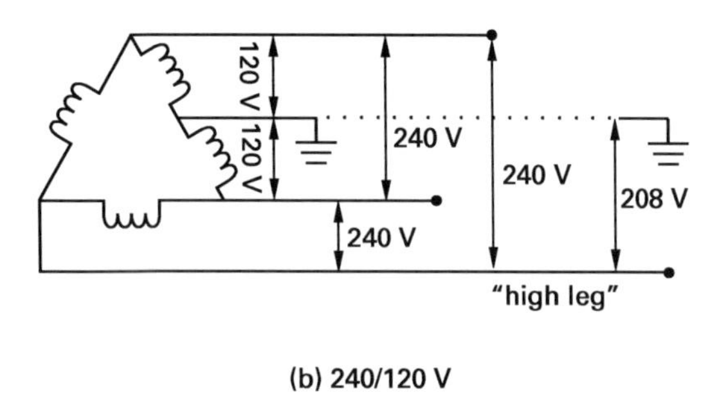

(b) 240/120 V

Figure 56.4 *Ground-Fault Circuit Interrupter (GFCI)*

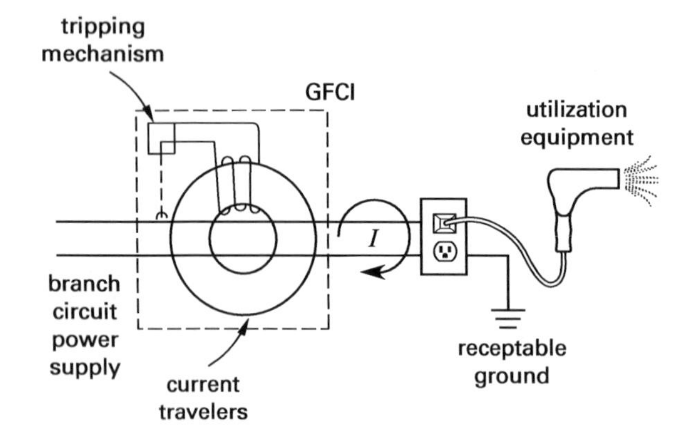

EPRM Chapter 57
National Electrical Safety Code

1. OVERVIEW

Figure 57.1 *NESC Areas of Concern*

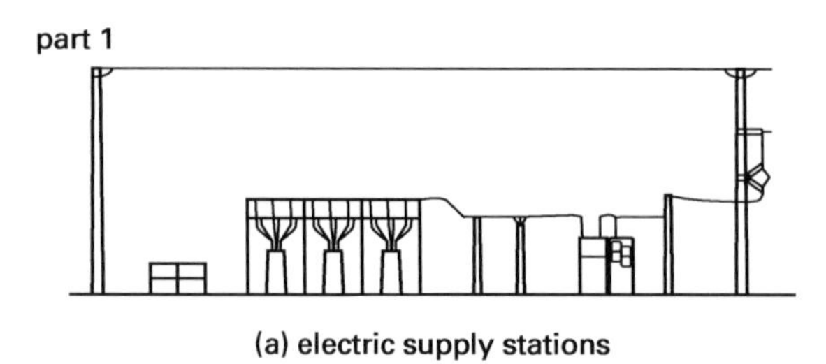

(a) electric supply stations

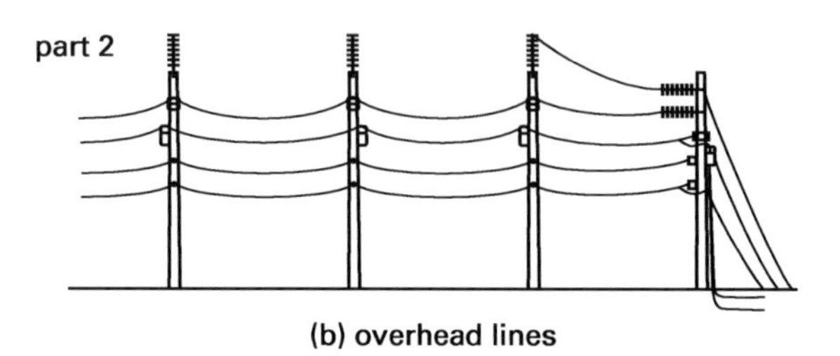

(b) overhead lines

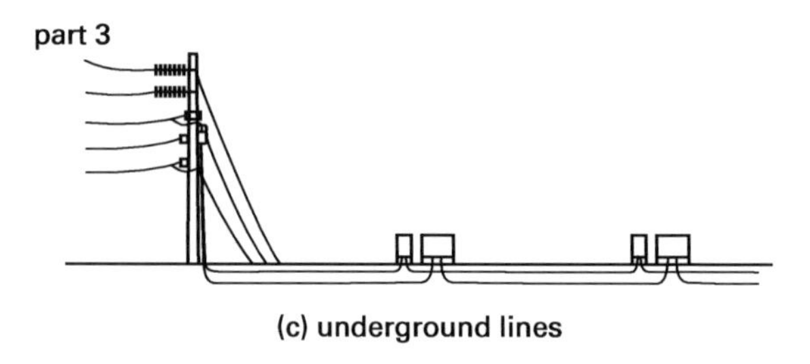

(c) underground lines

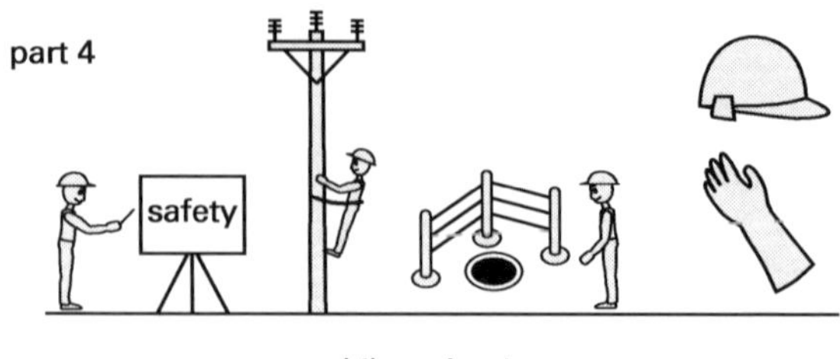

(d) work rules

2. INTRODUCTION

Figure 57.2 *NESC Coverage*

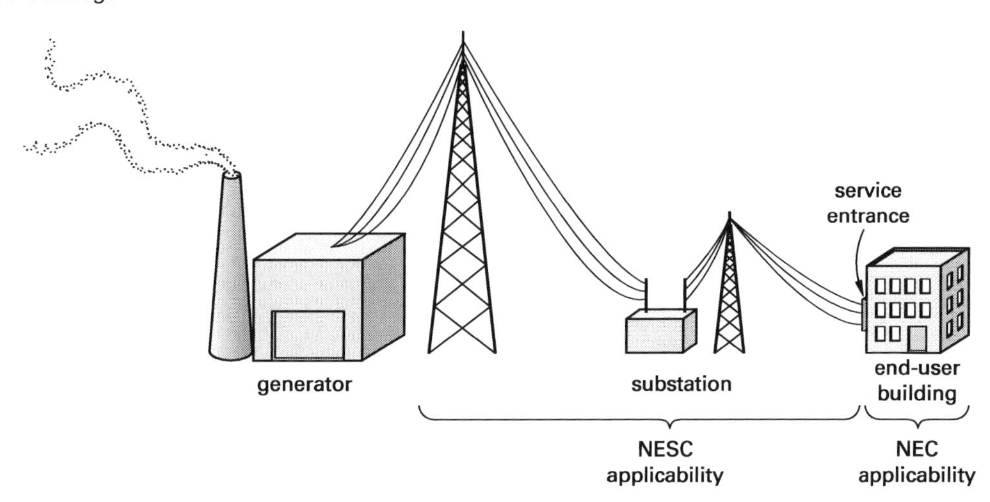

Figure 57.3 *NEC/NESC Division of Responsibility*

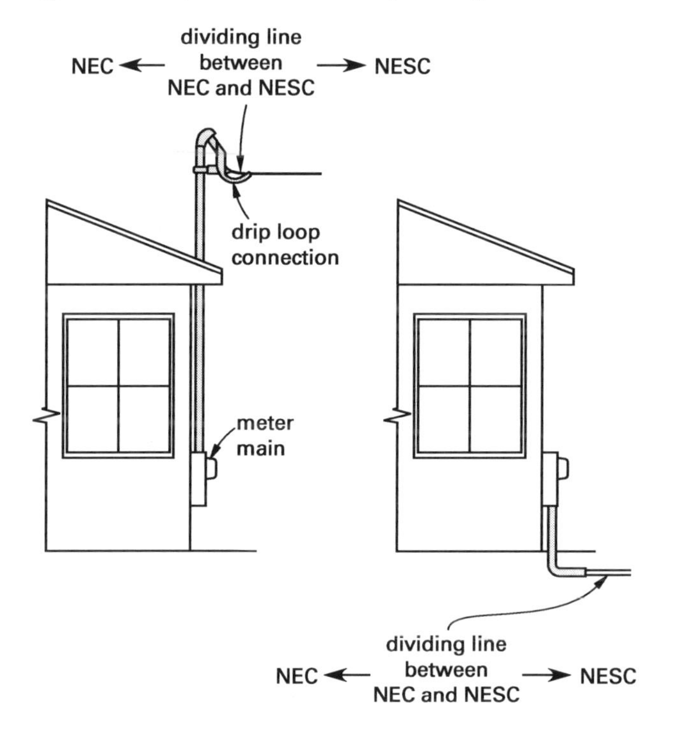

5. GROUNDING

Figure 57.4 *Allowable Ground Connections (≤ 750 V)*

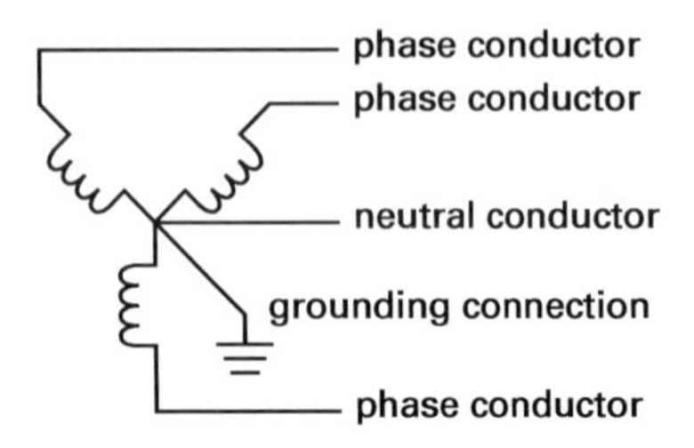

wye connected, 3-phase, 4-wire system
(e.g., 120/208 V, 3ϕ (phase), 4 W, and 277/480 V,
3ϕ (phase), 4 W)

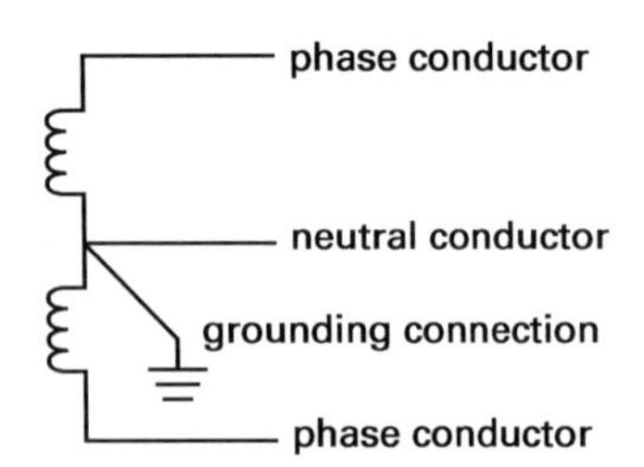

single-phase, 3-wire system
(e.g., 120/240 V, 1ϕ (phase), 3 W)

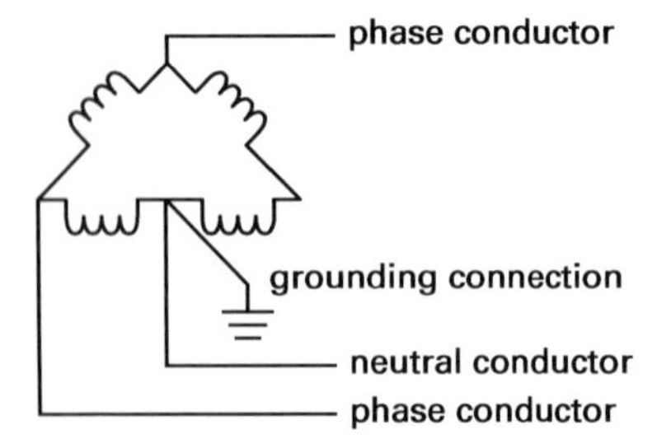

single-phase, 3-wire system on a delta connected 3-phase,
4-wire system (e.g., 120/240 V, 1ϕ (phase), 3 W)

Professional Practice

EPRM Chapter 58
Engineering Economic Analysis

11. SINGLE-PAYMENT EQUIVALENCE

$$F = P(1+i)^n \qquad 58.2$$

$$P = F(1+i)^{-n} = \frac{F}{(1+i)^n} \qquad 58.3$$

12. STANDARD CASH FLOW FACTORS AND SYMBOLS

(See Table 58.1.)

24. CHOICE OF ALTERNATIVES: COMPARING ONE ALTERNATIVE WITH ANOTHER ALTERNATIVE

Capitalized Cost Method

$$\text{capitalized cost} = \text{initial cost} + \frac{\text{annual costs}}{i} \qquad 58.19$$

$$\begin{aligned}\text{capitalized cost} &= \text{initial cost} + \frac{\text{EAA}}{i} \\ &= \text{initial cost} + \text{present worth of all expenses}\end{aligned} \qquad 58.20$$

Table 58.1 *Discount Factors for Discrete Compounding*

factor name	converts	symbol	formula
single payment compound amount	P to F	$(F/P, i\%, n)$	$(1+i)^n$
single payment present worth	F to P	$(P/F, i\%, n)$	$(1+i)^{-n}$
uniform series sinking fund	F to A	$(A/F, i\%, n)$	$\frac{i}{(1+i)^n - 1}$
capital recovery	P to A	$(A/P, i\%, n)$	$\frac{i(1+i)^n}{(1+i)^n - 1}$
uniform series compound amount	A to F	$(F/A, i\%, n)$	$\frac{(1+i)^n - 1}{i}$
uniform series present worth	A to P	$(P/A, i\%, n)$	$\frac{(1+i)^n - 1}{i(1+i)^n}$
uniform gradient present worth	G to P	$(P/G, i\%, n)$	$\frac{(1+i)^n - 1}{i^2(1+i)^n} - \frac{n}{i(1+i)^n}$
uniform gradient future worth	G to F	$(F/G, i\%, n)$	$\frac{(1+i)^n - 1}{i^2} - \frac{n}{i}$
uniform gradient uniform series	G to A	$(A/G, i\%, n)$	$\frac{1}{i} - \frac{n}{(1+i)^n - 1}$

25. CHOICE OF ALTERNATIVES: COMPARING AN ALTERNATIVE WITH A STANDARD

Benefit-Cost Ratio Method

$$B/C = \frac{\Delta \begin{matrix}\text{user}\\ \text{benefits}\end{matrix}}{\Delta \begin{matrix}\text{investment}\\ \text{cost}\end{matrix} + \Delta \,\text{maintenance} - \Delta \begin{matrix}\text{residual}\\ \text{value}\end{matrix}} \quad 58.21$$

26. RANKING MUTUALLY EXCLUSIVE MULTIPLE PROJECTS

$$\frac{B_2 - B_1}{C_2 - C_1} \geq 1 \quad \text{[alternative 2 superior]} \quad 58.22$$

30. TREATMENT OF SALVAGE VALUE IN REPLACEMENT STUDIES

$$\begin{aligned}\text{EUAC (defender)} &= \text{next year's maintenance costs} \\ &\quad + i(\text{current salvage value}) \\ &\quad + \text{current salvage} \\ &\quad - \text{next year's salvage}\end{aligned} \quad 58.23$$

35. DEPRECIATION BASIS OF AN ASSET

$$\text{depreciation basis} = C - S_n \quad 58.24$$

36. DEPRECIATION METHODS

Straight Line Method

$$D = \frac{C - S_n}{n} \quad 58.25$$

Sum-of-the-Years' Digits Method

$$T = \tfrac{1}{2}n(n+1) \quad 58.27$$

$$D_j = \frac{(C - S_n)(n - j + 1)}{T} \quad 58.28$$

Double Declining Balance Method

$$D_{\text{first year}} = \frac{2C}{n} \quad 58.29$$

$$D_j = \frac{2\left(C - \sum_{m=1}^{j-1} D_m\right)}{n} \quad 58.30$$

$$d = \frac{2}{n} \quad 58.31$$

$$D_j = dC(1-d)^{j-1} \quad 58.32$$

Statutory Depreciation Systems

$$D_j = C \times \text{factor} \quad 58.33$$

Table 58.4 *Representative ACRS and MACRS Depreciation Factors**

year j	recovery period, n: 3 yr ACRS/MACRS	5 yr ACRS/MACRS	10 yr ACRS/MACRS
1	0.25/0.3333	0.15/0.2000	0.08/0.100
2	0.38/0.4445	0.22/0.3200	0.14/0.180
3	0.37/0.1481	0.21/0.1920	0.12/0.1440
4	0/0.0741	0.21/0.1152	0.10/0.1152
5		0.21/0.1152	0.10/0.0922
6		0/0.0576	0.10/0.0737
7			0.09/0.0655
8			0.09/0.0655
9			0.09/0.0656
10			0.09/0.0655
11			0/0.0328

*MACRS values are for the "half-year" convention. This table gives typical values only. Because these factors are subject to continuing revision, they should not be used without consulting an accounting professional.

Production or Service Output Method

$$D_j = (C - S_n)\left(\frac{\text{actual output in year } j}{\text{estimated lifetime output}}\right) \quad 58.34$$

Sinking Fund Method

$$D_j = (C - S_n)(A/F, i\%, n)(F/P, i\%, j-1) \quad 58.35$$

38. BOOK VALUE

For the straight line depreciation method, the book value at the end of the jth year, after the jth depreciation deduction has been made, is

$$\text{BV}_j = C - \frac{j(C - S_n)}{n} = C - jD \quad 58.39$$

For the sum-of-the-years' digits method, the book value is

$$\text{BV}_j = (C - S_n)\left(1 - \frac{j(2n+1-j)}{n(n+1)}\right) + S_n \quad 58.40$$

For the declining balance method, including double declining balance, the book value is

$$\text{BV}_j = C(1-d)^j \quad 58.41$$

For the sinking fund method, the book value is calculated directly as

$$\text{BV}_j = C - (C - S_n)(A/F, i\%, n)(F/A, i\%, j) \quad 58.42$$

For any method by successive subtractions, the book value is

$$\text{BV}_j = C - \sum_{m=1}^{j} D_m \quad 58.43$$

41. BASIC INCOME TAX CONSIDERATIONS

$$t = s + f - sf \quad 58.44$$

45. RATE AND PERIOD CHANGES

$$\phi = \frac{r}{k} \quad 58.53$$

$$i = (1+\phi)^k - 1 = \left(1 + \frac{r}{k}\right)^k - 1 \quad 58.54$$

47. PROBABILISTIC PROBLEMS

$$\mathcal{E}\{\text{cost}\} = p_1(\text{cost } 1) + p_2(\text{cost } 2) + \cdots \quad 58.55$$

50. ACCOUNTING PRINCIPLES

$$\text{assets} = \text{liability} + \text{owner's equity} \quad 58.56$$

53. BREAK-EVEN ANALYSIS

$$C = f + aQ \quad 58.69$$

$$R = pQ \quad 58.70$$

$$Q^* = \frac{f}{p-a} \quad 58.71$$

Figure 58.13 *Break-Even Quantity*

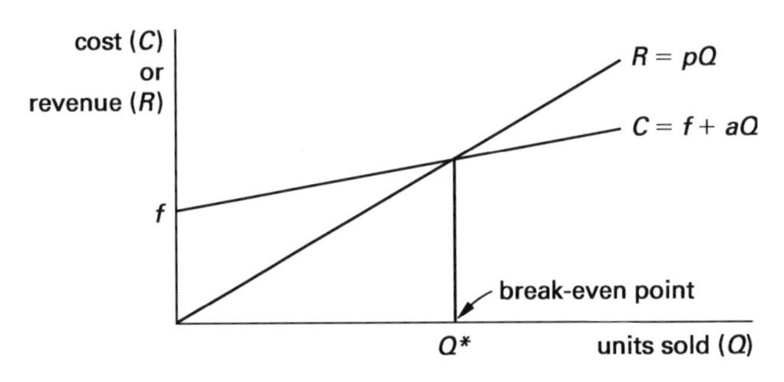

56. INFLATION

$$i' = i + e + ie \quad 58.73$$

58. FORECASTING

Forecasts by Moving Averages

$$F_{t+1} = \frac{1}{n} \sum_{m=t+1-n}^{t} D_m \quad 58.87$$

Forecasts by Exponentially Weighted Averages

$$F_{t+1} = \alpha D_t + (1-\alpha)F_t \quad 58.88$$

59. LEARNING CURVES

$$T_n = T_1 n^{-b} \quad 58.89$$

$$\int_{n_1}^{n_2} T_n \, dn \approx \left(\frac{T_1}{1-b}\right)\left(\left(n_2 + \tfrac{1}{2}\right)^{1-b} - \left(n_1 - \tfrac{1}{2}\right)^{1-b}\right) \quad 58.90$$

$$T_{\text{ave}} = \frac{\int_{n_1}^{n_2} T_n \, dn}{n_2 - n_1 + 1} \quad 58.91$$

$$b = \frac{-\log_{10} R}{\log_{10}(2)} = \frac{-\log_{10} R}{0.301} \quad 58.92$$

60. ECONOMIC ORDER QUANTITY

$$t^* = \frac{Q}{a} \quad 58.93$$

$$H = \tfrac{1}{2} Q h t^* = \frac{Q^2 h}{2a} \quad 58.94$$

$$C_t = \frac{aK}{Q} + \frac{hQ}{2} \quad 58.95$$

$$Q^* = \sqrt{\frac{2aK}{h}} \quad 58.96$$

$$t^* = \frac{Q^*}{a} \quad 58.97$$

Figure 58.18 *Inventory with Instantaneous Reorder*

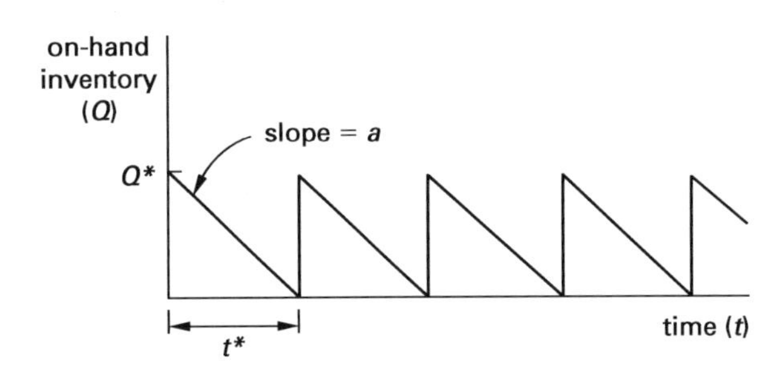

Index

D

E

F

G

H

I

J

K

L

Q

R

S

T

U

V

W

Y

Z